W0245750

PERSPECTIVES
IN
NEURAL COMPUTING

J.G. Taylor and C.L.T. Mannion (Eds.)

COUPLED OSCILLATING NEURONS

Springer-Verlag
London Berlin Heidelberg New York
Paris Tokyo Hong Kong
Barcelona Budapest

J. G. Taylor, BA, BSc, MA, PhD, FInstP
Department of Mathematics, King's College, Strand, London WC2R
2LS, UK

C. L. T. Mannion, BSc, PhD, MInstP
Department of Electrical Engineering, University of Surrey, Guildford,
Surrey, GU2 5XH, UK

Series Editors

J. G. Taylor, BA, BSc, MA, PhD, FInstP
Department of Mathematics, King's College, Strand, London WC2R
2LS, UK

C. L. T. Mannion, BSc, PhD, MInstP
Department of Electrical Engineering, University of Surrey, Guildford,
Surrey, GU2 5XH, UK

ISBN-13:978-3-540-19744-7   e-ISBN-13:978-1-4471-1965-4
DOI: 10.1007/978-1-4471-1965-4

British Library Cataloguing in Publication Data
Coupled Oscillating Neurons
(Perspectives in Neural Computing Series)
I. Taylor, John II. Mannion, C.L.T. III. Series
006.3
ISBN-13:978-3-540-19744-7

Library of Congress Cataloging-in-Publication Data
Coupled oscillating neurons / J.G. Taylor and C.L.T. Mannion, eds.
    p. cm. – (Perspectives in neural computing)
ISBN-13:978-3-540-19744-7 (Springer-Verlag New York Berlin Heidelberg)

1. Neurons. 2. Neural networks (Computer science) 3. Nonlinear oscillators.
I. Taylor, John Gerald. 1931– . II. Mannion, C.L.T. III. Series.
QP363.3.C68 1992   92-3824
591.1'88–dc20   CIP

Apart from any fair dealing for the purposes of research or private study, or criticism
or review, as permitted under the Copyright, Designs and Patents Act 1988, this
publication may only be reproduced, stored or transmitted, in any form or by any
means, with the prior permission in writing of the publishers, or in the case of
reprographic reproduction in accordance with the terms of licences issued by the
Copyright Licensing Agency. Enquiries concerning reproduction outside those terms
should be sent to the publishers.

© Springer-Verlag London Limited 1992

The use of registered names, trademarks etc. in this publication does not imply, even
in the absence of a specific statement, that such names are exempt from the relevant
laws and regulations and therefore free for general use.

The publisher makes no representation, express or implied, with regard to the
accuracy of the information contained in this book and cannot accept any legal
responsibility or liability for any errors or omissions that may be made.

34/3830-543210 Printed on acid-free paper

PREFACE

This volume consists of proceedings of the one-day conference on "Coupled Oscillating Neurons" held at King's College, London on December 13th, 1990.

The subject is currently of increasing interest to neurophysiologists, neural network researchers, applied mathematicians and physicists. The papers attempt to cover the major areas of the subject, as the titles indicate. It is hoped that the appearance of the papers (some of which have been updated since their original presentation) indicates why the subject is becoming of great excitement.

A better understanding of coupled oscillating neurons may well hold the key to a clearer appreciation of the manner in which neural networks composed of such elements can control complex behaviour from the heart to consciousness.

December 1991   J.G. Taylor
King's College, London   C.L.T. Mannion

CONTENTS

CONTRIBUTORS

Ashwin, P.B.
Mathematics Institute, University of Warwick, Coventry, CV4 7AL, UK

Brown, H.F.
University Laboratory of Physiology, Parks Road, Oxford, OX1 3PT, UK

Cotterill, R.M.J.
Division of Molecular Biophysics, The Technical University of Denmark,
Building 307, DK-2800 Lyngby, Denmark

Denyer, J.C.
University Laboratory of Physiology, Parks Road, Oxford, OX1 3PT, UK

Hindmarsh, J.L.
School of Mathematics, Senghennydd Road, PO Box No 915, Cardiff,
CF2 4AG, UK

Holden, A.V.
Department of Physiology and Centre for Nonlinear Studies, University
of Leeds, Leeds, LS2 9JT, UK

Hyde, J.
Department of Physiology and Centre for Nonlinear Studies, University
of Leeds, Leeds, LS2 9JT, UK

Kimball, A.
Department of Physiology and Army High Performance Computer
Center, University of Minnesota, USA

Mannion, C.L.T.
Department of Electrical Engineering, University of Surrey, Guildford,
Surrey, GU2 5XH, UK

Muhamad, M.A.
Department of Physiology and Centre for Nonlinear Studies, University
of Leeds, Leeds, LS2 9JT, UK

Nielsen, C.
Division of Molecular Biophysics, The Technical University of Denmark,
Building 307, DK-2800 Lyngby, Denmark

Noble, D.
University Laboratory of Physiology, Parks Road, Oxford, OX1 3PT, UK

Rose, R.M.
Department of Physiology, Senghennyd Road, PO Box No 915,
Cardiff, CF2 4AG, UK

Stewart, I.
Mathematics Institute, University of Warwick, Coventry, CV4 7AL, UK

Taylor, J.G.
Department of Mathematics, King's College London, Strand, London,
WC2R 2LS, UK

Winslow, R.
Department of Physiology and Army High Performance Computer
Center, University of Minnesota, USA

Zhang, H.G.
Department of Physiology and Centre for Nonlinear Studies, University
of Leeds, Leeds, LS2 9JT, UK

Introduction to Nonlinear Oscillators

Ian Stewart

Mathematics Institute
University of Warwick
Coventry CV4 7AL
England

Abstract

We discuss some of the basic mathematical theory of nonlinear oscillators, both autonomous and forced. This includes the birth and death of periodic oscillations in Hopf and homoclinic bifurcations, and the technique of Poincaré sections, which reduces dynamics near a periodic oscillation to that of a discrete mapping. The kicked rotator, itself a discrete system, is used to introduce the notions of frequency-locking, Arnold tongues, and the devil's staircase. General features of symmetry-breaking in coupled systems of identical oscillators are described, and applications to Central Pattern Generator models of animal gaits are sketched.

1 Nonlinear Oscillators

1.1 Introduction and Examples

An oscillator is a dynamical system that behaves in a (more or less) periodic fashion. Biological systems, in particular, are often oscillators, see Glass and Mackey [14]. Examples include electrical activity in the brain (EEG records, see Fig.1); the heartbeat (ECG records, Fig.2); and the production of white cells in the blood (Fig.3).

Mathematically, two basic types of oscillator should be distinguished:

Unforced Oscillator

$$\frac{dx}{dt} + f(x) = 0 \qquad (x \in \mathbb{R}^n, \ f{:}\mathbb{R}^n \to \mathbb{R}^n) \tag{1}$$

Forced Oscillator

$$\frac{dx}{dt} + f(x) = A\varphi(t) \quad (x \in \mathbb{R}^n, \ f{:}\mathbb{R}^n \to \mathbb{R}^n, \ \varphi{:}\mathbb{R} \to \mathbb{R}^n) \tag{2}$$

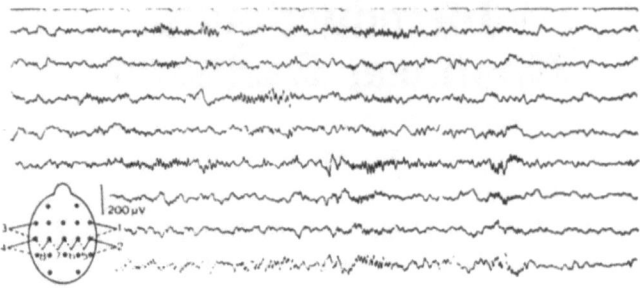

Fig.1 Electroencephalograph record of brain activity exhibits a mixture of approximately periodic oscillations. [From Glass and Mackey [14].]

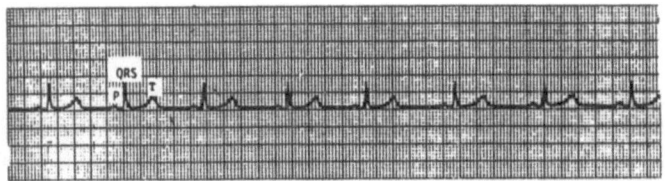

Fig.2 Electrocardiograph record of a heartbeat. [From Glass and Mackey [14].]

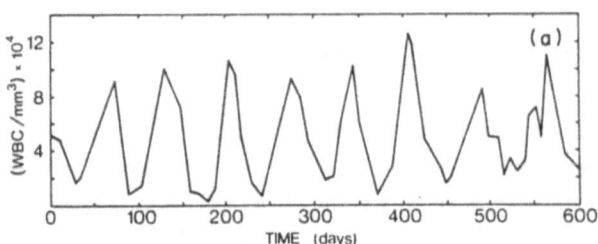

Fig.3 In some diseases, the white blood cell count oscillates. [From Glass and Mackey [14].]

where φ is T-*periodic*, that is, $\varphi(t+T) = \varphi(t)$ for all $t \in \mathbb{R}$. Throughout this paper we assume for convenience that the vector field f and the forcing function φ are infinitely differentiable.

An unforced oscillator is a free-running system that tends to oscillate of its own accord. A forced oscillator is driven by a signal $\varphi(t)$ which, effectively, is produced by a second free-running oscillator.

The most familiar oscillator is the *simple harmonic oscillator*, whose equation

$$\frac{d^2x}{dt^2} + k^2 x$$

can be rewritten as a first order system

$$\frac{d}{dt}\begin{pmatrix} x \\ y \end{pmatrix} = \begin{pmatrix} 0 & -k \\ k & 0 \end{pmatrix}\begin{pmatrix} x \\ y \end{pmatrix} \tag{3}$$

by setting $x = \frac{1}{k}\frac{dy}{dt}$. This has the well-known explicit solution

$$y = A \cos kt + B \sin kt.$$

The simple harmonic oscillator is a linear oscillator, because the function f that arises when we write (3) in the form (1) is linear in x and y. The behaviour of linear oscillators is relatively straightforward. In contrast, *nonlinear* oscillators, in which f is not a linear function, behave in surprising and often counter-intuitive ways. Linearity is an extremely special property, and most oscillators encountered in nature are better modelled by nonlinear oscillators, so we concentrate on the nonlinear case. Archetypal model oscillators are the *Van der Pol oscillator*, with equation,

$$\frac{dx}{dt} = y - \alpha(\frac{x^3}{3} - x)$$

$$\frac{dy}{dt} = -x, \tag{4}$$

and the *Duffing oscillator*

$$\frac{dx}{dt} = y - y^3 - \delta x + \gamma \cos \omega t$$

$$\frac{dy}{dt} = x. \tag{5}$$

The Van der Pol oscillator is unforced, the Duffing forced. To demonstrate that nonlinear oscillators are not straightforward, we need do no more than reproduce a famous figure obtainbed by Hayashi [20] using an analogue computer (Fig.4). This shows how the state of a Duffing oscillator depends upon the initial values of (x,y). For general information about the dynamics of nonlinear oscillators see Arrowsmith and Place [5],Guckenheimer and Holmes [19], and Jackson [22].

4

1.2 Birth and Death of an Oscillator

How does an oscillation begin? The commonest scenario is called *Hopf bifurcation*. Here the dynamics of the oscillator changes as some parameter λ is varied, and a stable steady state becomes unstable, 'throwing off' a limit cycle. We can visualise this process geometrically using a *phase portrait* (Fig.5), which represents the solutions as curves in the *phase space* definedby the dynamical variables x of the system. A closed curve in a phase portrait represents a periodic oscillation.

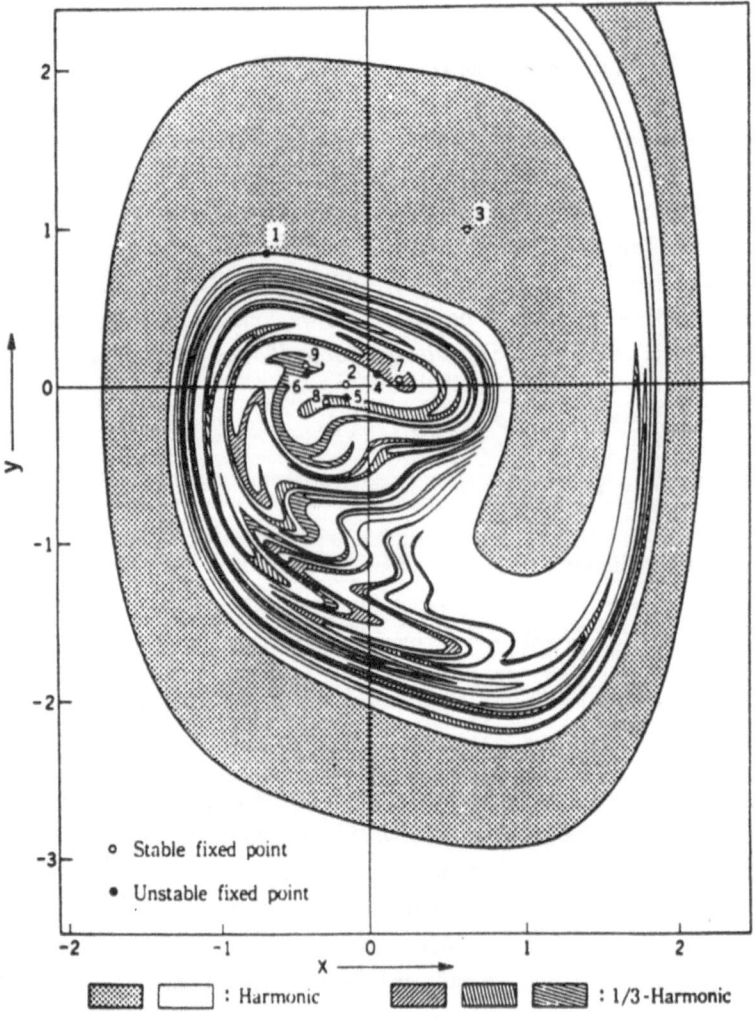

Fig.4 How the state of a Duffing oscillator varies with initial conditions. Five oscillations (two harmonic and three subharmonic resonances) are shown. [From Hayashi [20].]

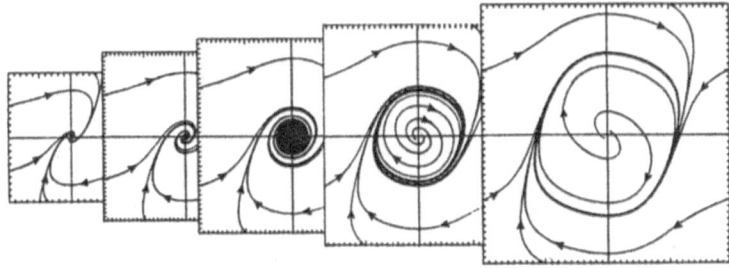

Fig.5 Hopf bifurcation to a limit cycle. [From Thompson and Stewart [30].]

The same phenomenon is shown schematically in Fig.6a, known as a *bifurcation diagram*. Here the amplitude of the oscillation increases continuously from zero. A slight variation (Fig.6b) leads to the abrupt appearance of a periodic oscillation of finite amplitude. In fact, as the bifurcation parameter λ increases, first two limit cycles, one stable and one unstable, are created simultaneously from nothing; subsequently the unstable cycle merges with the stable steady state in a reverse Hopf bifurcation to create an unstable steady state.

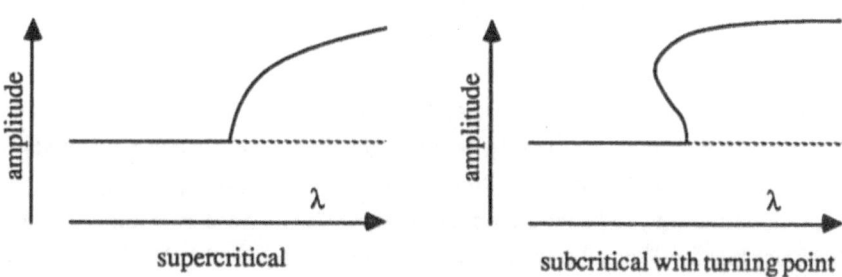

Fig.6 Supercritical and subcritical Hopf bifurcations.

Hopf bifurcations cause limit cycles to be born, but how do limit cycles die? Obviously one way is by reversing the sequence of changes in a Hopf bifurcation; but another common phenomenon is a *homoclinic bifurcation* or *blue sky catastrophe* (Fig.7) in which the limit cycle runs into a saddle point. The period of the limit cycle tends to infinity at a homoclinic bifurcation. See Abraham and Marsden [1], Abraham and Shaw [2].

1.3 Poincaré Sections

A technique for reducing the dynamics of a continuous n-dimensional system to that of a discrete $(n-1)$-dimensional system was introduced by Henri Poincaré. In its simplest form it applies to the dynamics near a limit cycle L (for example, this is

6

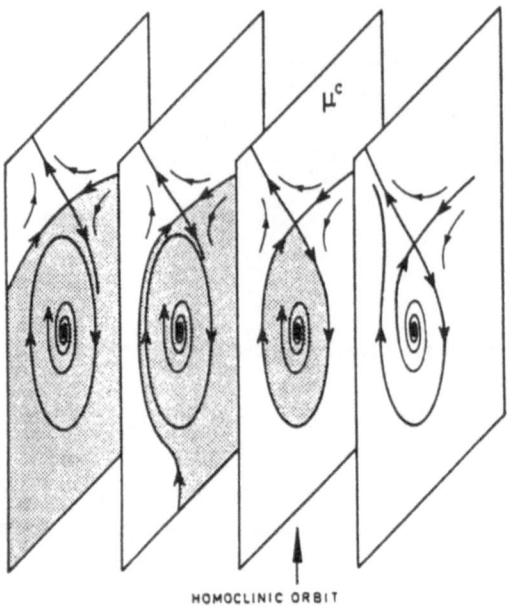

HOMOCLINIC ORBIT

Fig.7 Homoclinic bifurcation. [From Thompson and Stewart [30].]

important in the study of stability of the limit cycle). Suppose that the dimension of phase space is n. The idea is to choose an $(n\text{-}1)$-dimensional hypersurface Σ that cuts L transversely. It then cuts all nearby orbits of the dynamical system transversely as well. We call Σ a *Poincaré section*. Suppose that L meets Σ in a point x_0. For each point $x \in \Sigma$, define $\varphi(x)$ to be the first point at which the forward trajectory of x under the dynamics hits Σ. We call φ the *Poincaré mapping* associated to Σ. By continuity, $\varphi(x)$ is defined for x in some neighbourhood U of x_0, and $\varphi : U \to \Sigma$. Poincaré

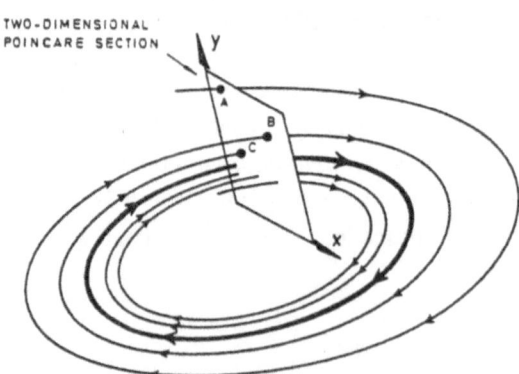

Fig.8 A Poincaré section and its Poincaré mapping. [From Thompson and Stewart [30].]

proved that φ is a local diffeomorphism; that is, it is infinitely differentiable and invertible on some neighbourhood of x_0. Since L is a cycle, $\varphi(x_0) = x_0$; that is, x_0 is a fixed point of φ. See Fig.8.

Iteration of φ describes that part of the dynamics that moves x transversely to L. For example, L is a stable limit cycle if and only if x_0 is a stable fixed point of φ. The technique of Poincaré sections is especially useful when L is unstable: for example, bifurcations of L, or nearby chaos, may be detected.

A variant of this technique can be used for forced oscillators. Now Σ is the whole of phase space, and $\varphi(x)$ is the point to which x moves under the dynamics after a time of one forcing period (Fig.9). Another name for this technique is *stroboscopic sampling*.

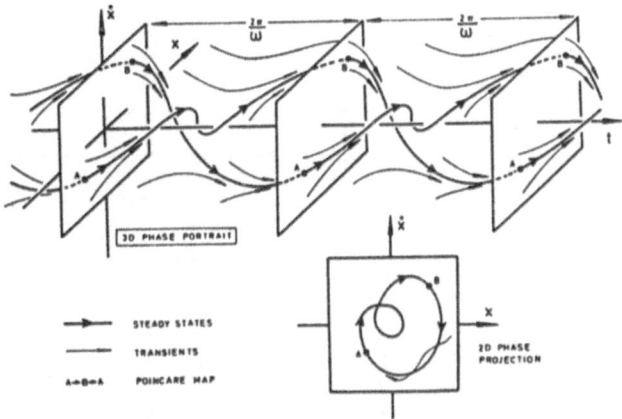

Fig.9 Poincaré section for a forced system. [From Thompson and Stewart [30].]

1.4 Examples

The Duffing equation (5) possesses chaotic solutions. The corresponding Poincaré section shows that these have considerable structure, see Fig.10.

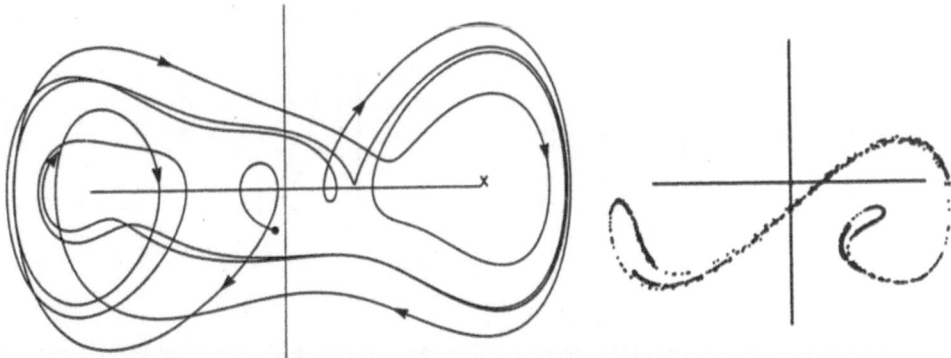

Fig.10 A chaotic trajectory of the Duffing equation and its Poincaré mapping. [From Thompson and Stewart [30].]

8

Poincaré mappings are especially useful when investigating experimental oscillations, since they can reveal deterministic structure in the same manner. We give two examples. The first occurs in an investigation of the Van der Pol - Duffing oscillator, whose circuit is shown in Fig.12. Chaotic oscillations of this system, found experimentally by Castro [10], and the corresponding Poincaré sections and mappings, are shown in Fig.11. A similar analysis of oscillations in the Belousov-Zhabotinskii reaction, due to Roux *et al.* [25], is shown in Fig.13.

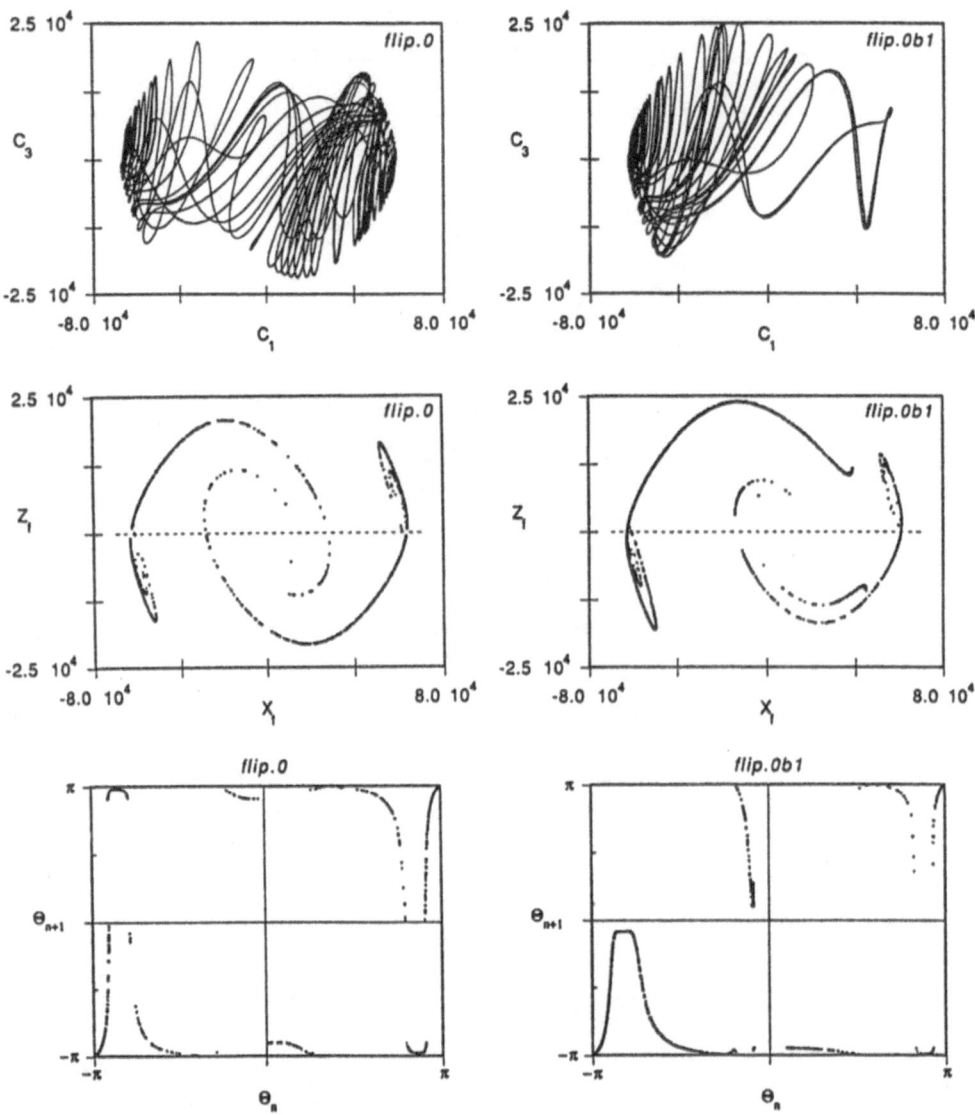

Fig.11 (a) Chaotic trajectories of the Van der pol -Duffing oscillator, (b) their Poincaré sections, and (c) their Poincaré mappings. [From Castro [10].]

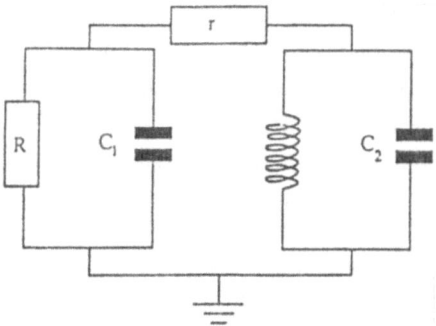

Fig.12 Circuit diagram of the Van der pol -Duffing oscillator.

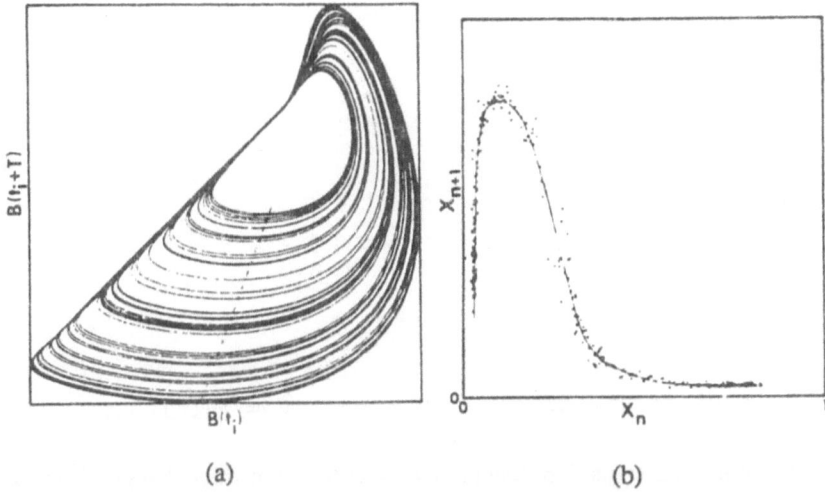

(a)   (b)

Fig.13 (a) A chaotic trajectory of the Belousov-Zhabotinskii reaction, (b) its Poincaré mapping. [From Roux *et al.* [25].]

2   Frequency-Locking

We now turn to forced oscillators, and investigate the phenomenon of *frequency-locking*, which occurs when the frequency of the oscillator is a rational multiple of the forcing frequency.

To illustrate another theoretical technique, we now model the forced oscillator by a discrete system, often called the *kicked rotator*:

$$\varphi_{t+1} = (\varphi_t - \frac{K}{2\pi}\sin(2\pi\varphi_t) + \Omega \tag{6}$$

Here the state at time t, an integer, is represented by φ_t, which takes values between 0

10

and 1, with 0 and 1 being identified to make the phase space topologically equivalent to a circle. The term Ω represents regular forcing, and the 'natural' nonlinear oscillations of the system are given by setting $\Omega = 0$. Physically, we can think of φ as the phase of a continuous oscillator which is 'kicked' by an amount Ω at regular intervals of time, and is assumed to relax rapidly back to its original limit cycle.

Frequency-locking is an important feature of the kicked rotator. For each rational number p/q there exists a wedge-shaped region of (Ω,K)-space in which the ratio of the frequency of oscillation to the forcing frequency is p/q, see Fig.14. These regions, known as *Arnold tongues*, overlap in a complicated manner which eventually leads to chaos. See Arnold [4], Schuster [27].

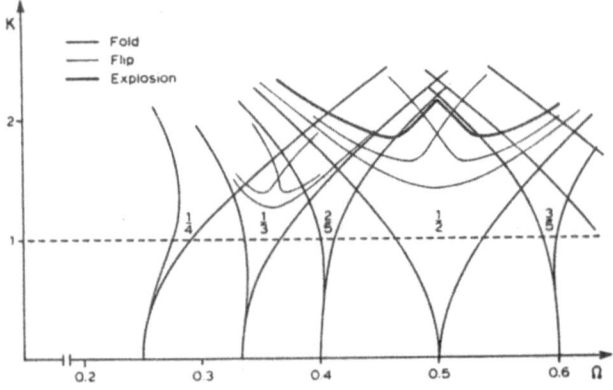

Fig.14 Arnold tongues in the kicked rotator. [From Thompson and Stewart [30].]

An important transition occurs at $K = 1$: here the map changes from being invertible to being non-invertible. A graph of the observed frequency of a stable oscillation against the forcing frequency Ω takes the form of a *Devil's staircase* or Cantor function, composed of almost everywhere horizontal lines, Fig.15. The graph is *self-similar*: small pieces, when magnified, resemble the whole.

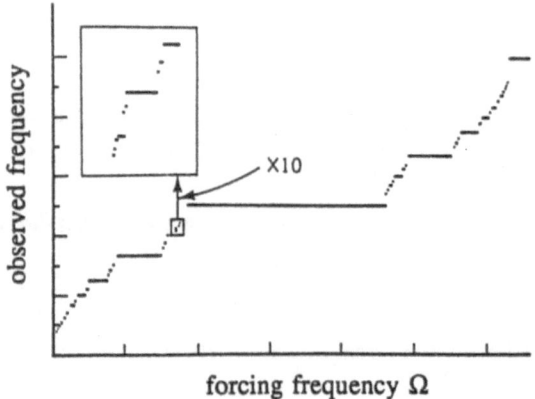

forcing frequency Ω

Fig.15 The Devil's Staircase. [From Jackson [22].]

3   Coupled Identical Oscillators

In this final section we describe some recent results on analogues of Hopf bifurcation for systems of coupled *identical* oscillators, and illustrate them with a speculative but striking application to the distinct patterns of leg-movement that animals employ at different speeds of locomotion, known as *gaits*. For example humans walk, run, leap or hop; and horses walk, trot, canter, or gallop. We model gait patterns as symmetry-breaking oscillations of a network composed of a number of identical neural subcircuits. The results presented here are a summary of those obtained in Collins and Stewart [11], and complement those of Schöner *et al.* [26], which approaches the same problem from the point of view of synergetics

3.1 Symmetries of Animal Gaits

Hildebrand [21] emphasised that most gaits possess a degree of symmetry. For example, in the *bound* both front legs move together and both back legs move together (Fig.16), a gait with bilateral symmetry. The symmetries often involve phase shifts: for example the left half of an animal often follows the same sequence of movements as the right half, but half a period out of phase, as in Fig.17. This is an instance of *symmetry-breaking*, see Golubitsky *et al.* [17]: the gait of a bilaterally symmetric animal can fail to be bilaterally symmetric. However, such a gait has its own symmetry: 'interchange left and right sides *and* shift phase by half a period'.

Fig.16 The bound of the long-tailed Siberian souslik. [From Gambaryan [12].]

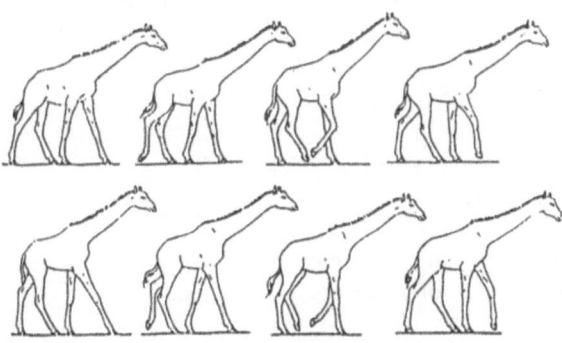

Fig.17 Slow pace-like walk of giraffe. [From Gambaryan [12].]

Our main interest is the gaits of quadrupeds, but the ideas are most easily grasped in the simpler case of a biped.   We therefore begin by listing the standard gaits of bipeds and quadrupeds.

Bipedal Gaits
Hop. The two limbs move in phase.
Walk, run. The two limbs move half a period out of phase.

Quadrupedal Gaits
Walk.   Successive legs move a quarter period out of phase in a figure-8 rotating wave.
Trot.   Diagonal pairs of legs move together and in phase. One pair is half a period out of phase with the other pair.
Pace. The left legs move together and in phase. The right legs move together, half a
        period out of phase with the left legs.
Canter.   Right front/left back legs move together and in phase. Left front and right
        back legs move half a period out of phase with another and out of phase with
        the strongly coupled diagonal pair.
Transverse Gallop. The two left legs are half a period of phase, and so are the two
        right legs.   There is a somewhat arbitrary phase lag between left and right.
Rotary Gallop.   Similar to transverse gallop except that now diagonal legs are half a
        period out of phase with one another.
Bound. The front legs move together and in phase. The back legs move together, half a period out of phase with the front pair.
Pronk.   All four legs move at the same time.
The phase relationships in these gaits are summarised in Fig.18.

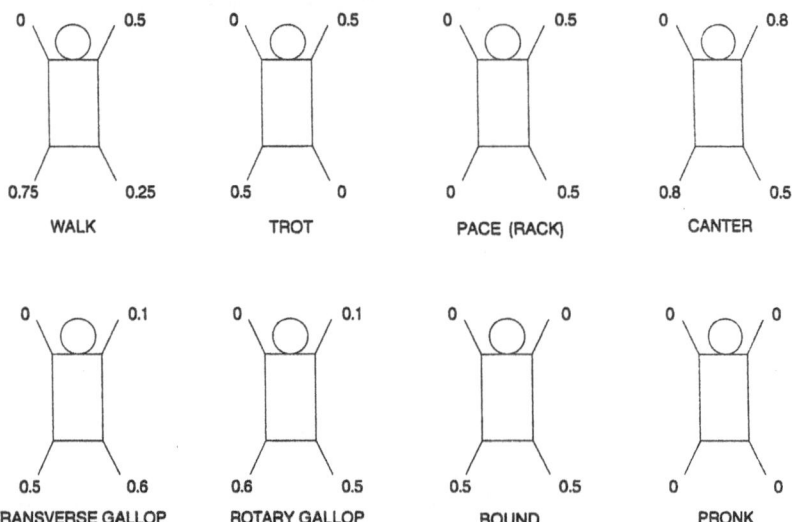

Fig.18 Phase relations in quadruped gaits.

3.2 Hopf Bifurcation in Coupled Systems

An animal or a CPG is not capable of changing its symmetry — so how can gaits with different symmetries arise? Why do the gaits change their symmetries depending on speed? Such phenomena fit neatly into the general pattern of symmetry-breaking bifurcation in symmetric nonlinear dynamical systems, and in particular in coupled oscillator systems.

The onset of oscillations by Hopf bifurcation has already been discussed in §1.2. There is an analogue of the Hopf bifurcation theorem for symmetric dynamical systems. Its statement involves technicalities from group represenatation theory, which we mention in order to illustrate the nature of the mathematical context: see the references cited for a full explanation. Consider the dynamical system

$$\frac{dx}{dt} + f(x, \lambda) = 0 \qquad (x \in \mathbb{R}^n, f{:}\mathbb{R}^n \to \mathbb{R}^n, \lambda \in \mathbb{R}). \qquad (7)$$

We call λ the *bifurcation parameter*. For convenience assume that $f(0,\lambda) \equiv, 0$, so that $x = 0$ is always a steady state solution of (7). Let $(df)_{(x,\lambda)}$ be the Jacobian matrix $[\partial f_i / \partial f_j]$ evaluated at (x,λ). Suppose that f commutes with the action of a compact Lie group Γ on $\mathbb{R}^n$. For Hopf bifurcation to occur, $(df)_{(x,\lambda)}$ must have purely imaginary eigenvalues $\pm i\omega$ at some value a_0 of λ. Assume that the eigenvalues cross the imaginary axis with non-zero speed. Generically the corresponding real eigenspace of the derivative $L = (df)_{(x,\lambda_0)}$ is a Γ-*simple* representation; that is, of the form

$V \oplus V$ where V is absolutely irreducible, or

W where W is non-absolutely irreducible.

Assume this generic hypothesis, and assume without loss of generality (via centre manifold or Liapunov-Schmidt reduction) that $\mathbb{R}^n$ is the real eigenspace of L for eigenvalues $\pm i\omega$. Define an action of the circle group $S^1 = \mathbb{R}/\mathbb{Z}$ on $\mathbb{R}^n$ by

$\theta.x = e^{-2\pi\theta L}$.

If $x \in \mathbb{R}^n$ then its *isotropy subgroup* $\Sigma_x \subset \Gamma \times S^1$ is defined to be

$\Sigma_x = \{\gamma \in \Gamma \times S^1 \mid \gamma.x = x\}$.

If $\Sigma \subset \Gamma \times S^1$ then its *fixed-point space* is defined to be

$\text{Fix}(\Sigma) = \{x \in \mathbb{R}^n \mid \sigma.x = x \text{ for all } \sigma \in \Sigma\}$.

With these assumptions, we have the following result of Golubitsky and Stewart [15]:

Symmetric Hopf Bifurcation Theorem

Let Σ be an isotropy subgroup of $\Gamma \times S^1$ such that $\dim \text{Fix}(\Sigma) = 2$. *Then there exists a branch of periodic solutions to* (7) *with period near $2\pi/\omega$, having Σ as their group of spatio-temporal symmetries, where S^1 acts on a periodic solution by phase shift.*

This theorem reduces the question of the existence of symmetry-breaking oscillations to purely group-theoretic calculations. Informally it asserts that at a symmetric analogue of a Hopf bifurcation, one or more branches of periodic solutions, usually several, bifurcate. They may be distinguished by their symmetry groups $\Sigma \subset \Gamma \times S^1$. In particular a branch exists for every isotropy subgroup Σ that satisfies the condition dim Fix(Σ) = 2. For example, Fig.19 shows the generic patterns of oscillation for a ring of n coupled identical oscillators, a system whose symmetry group is the dihedral group $\mathbf{D}_n$. In these pictures it is assumed that $\mathbf{D}_n$ acts on the imaginary eigenspace in its standard representation: other oscillatory modes, corresponding to non-standard representations, are also possible, and produce different patterns. In particular for each divisor m of n, the oscillators may group into n/m phase-locked sets of m equally spaced oscillators, interacting like a ring of n/m oscillators. See Golubitsky and Stewart [15] and Golubitsky et al. [17] for details. Other sources for information on coupled identical oscillators include Ashwin [6], Ashwin et al. [7], Ashwin and Swift [8], and Swift [29]. For three non-identical oscillators, see Baesens et al. [9].

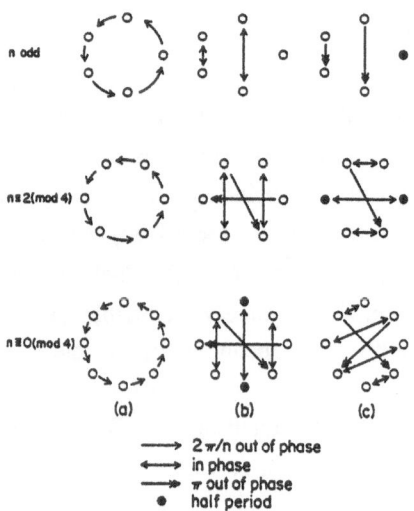

Fig.19   Some of the generic patterns for a ring of coupled identical oscillators. [From Golubitsky et al. [17].]

3.3 Central Pattern Generators

There are strong parallels between the generic nonlinear dynamics of coupled oscillators and gait patterns. A possible explanation lies in the concept of a Central Pattern Generator (CPG). This is a network of neurons, thought to control locomotion. A plausible reason for the observed parallels is that the CPG possesses symmetry, especially rectangular or square symmetry. The traditional approach to CPG architecture has been to hypothesise specific circuitry and analyse its dynamics.

Here we apply a model-independent approach to show that for each of a number of symmetry types of CPG, there is a 'universal' hierarchy of symmetry-breaking oscillation patterns, many of which correspond to actual gaits.

There is considerable evidence for the existence of a rhythm generator or a CPG: see the extensive discussion that follows Selverston [28]. Locomotion could be controlled by four (or more) distinct but coupled nonlinear oscillators. These might be 'central' in the sense that they occur in a small region of the animal, or 'distributed' about its body with longer neural interconnections; and for the purposes of this paper whenever we use the term CPG we include the 'distributed' option as well as the 'central'. Rand *et al.* [24] and Kopell [23] consider linear arrays of identical oscillators, or of pairs of identical oscillators. Getting [13] briefly describes a number of small neural networks.

3.4 Generic Oscillation Patterns

The Equivariant Hopf Bifurcation Theorem asserts the existence, under appropriate conditions, of a number of typical oscillation patterns for systems of two and four coupled nonlinear oscillators. We describe the results. For the four-oscillator system, we consider five possible cases, see Fig.20. Here there are either one or two types of oscillator, and up to three distinct types of coupling between them. We assume that the patterns of animal locomotion tend to reflect the organization of the underlying CPGs.

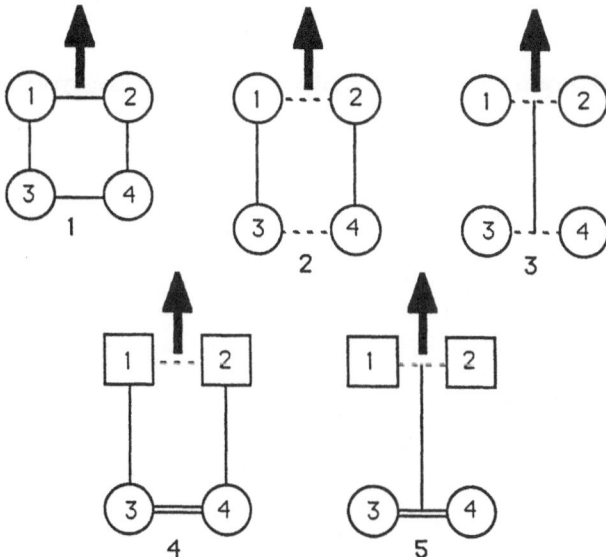

Fig.20 Graphical representation of five distinct symmetric patterns of coupled oscillators. Symbols o and □ indicate two distinct types of oscillator; lines ═══, - - - - -, and ─── indicate three distinct types of coupling. The arrow shows the direction in which the head is facing.

Two Oscillators

For two identical coupled oscillators there are two typical oscillation patterns:

(a)   The *in-phase* pattern: both oscillators have the same waveform.

(b)   The *out-of-phase* pattern: both oscillators have the same waveform except for a phase difference of half a period.

The gaits produced by such a CPG will be of two types. For the in-phase oscillation, both legs will move together; that is, the animal will perform a two-legged hopping motion, Fig.21. The out-of-phase motion could resemble the normal human walking gait, but a more exotic possibility with the same symmetry is shown in Fig.22. The picture shows only half a period: the 'final' position is the mirror image of the initial one.

Fig.21 Bound of kangaroo has in-phase left-right symmetry.   [From Gambaryan [12].].

Fig.22 Half-bound of Severtsov's jerboa has out-of-phase left-right symmetry.   Each foot hits the ground twice per period. [From Gambaryan [12].].

Four Oscillators

For the five coupled systems of Fig.20 the typical symmetries of periodic oscillations created by Hopf bifurcation are shown in Table 1. In some systems there may be additional patterns, but the existence of those listed is model-independent. The annotation BIFURCATE TOGETHER in Table 3 indicates states that occur together in a single Hopf bifurcation; that is, multiple branches of the bifurcation diagram that emerge from the same point. Here A and B represent oscillations, and $A+\varphi$ is A phase-shifted by a fraction φ of the period.

Table 1 shows that each type of network has its own particular set of 'natural' oscillation patterns. The most symmetric gaits (pronk, trot, bound, pace, walk) correspond precisely to patterns that occur in the table. The final gait listed for type **1** has the correct phase relations for a canter, but it is not clear (and seems unlikely) that a true canter involves the half-period property of waveform B. However, minor breaking of the square symmetry could destroy this property, leaving something closer to a canter. Michael Dellnitz has pointed out that systems with tetrahedral symmetry (four oscillators with fully symmetric coupling) exhibit oscillation patterns whose phase relations are similar to those that occur in the canter.

The rotary and transverse gallops are not represented in our list, although type **1** has two conjugate patterns, **1c** and **1d**, that are similar to the rotary gallop.

Among the twenty patterns listed for the first three types (all oscillators identical), only three do not seem to correspond to gaits listed in §2. All of these involve the 'half-period' condition.

The patterns for types **4** and **5** are all plausible for creatures whose front legs are very different from their rear legs. For example **5b** corresponds to a two-legged walk on hind legs, while the front legs move together; **4b** is essentially the normal human walking gait with B representing arm-movements and A leg-movements.

A similar analysis can be given for hexapod gaits, and again many of the observed gaits — such as the tripod gait of the cockroach, see Alexander [3] — correspond to natural patterns of symmetry-breaking oscillation in simple, symmetric CPGs.

Let us summarise our results. Networks of symmetrically coupled identical oscillators possess 'universal' patterns of phase-locked oscillations, many of which correspond to observed patterns of phase-locking in animal gaits. One way to develop these results would be to build detailed models of animal neurology or physiology, possessing the desired symmetries, and analyse the effect of various physical parameters, taking into account the known behaviour of symmetric networks. The role of our results in such an investigation would be to organize what behaviour should be looked for and suggest methods for finding it. Such models would involve adjustable parameters for such quantitites as the coupling between components (e.g. excitatory/inhibitory susceptibilities of individual neurons), or indeed their internal dynamics. The system will select possible patterns of oscillation from those available in general, according to the values of those parameters. This provides a natural mechanism for the control of gait transitions, and for subsequent learning.

Table 1 Typical Patterns for Coupled Nonlinear Oscillators

System	LF	RF	LH	RH	Mathematical Comments	Corresponding Gait
1 a	A	A	A	A		pronk
b	A	$A+\frac{1}{2}$	$A+\frac{1}{2}$	A		trot
c	A	$A+\frac{1}{4}$	$A+\frac{3}{4}$	$A+\frac{1}{2}$		similar to rotary gallop
d	A	$A+\frac{3}{4}$	$A+\frac{1}{4}$	$A+\frac{1}{2}$		similar to rotary gallop (opposite orientation)
e	A	A	$A+\frac{1}{2}$	$A+\frac{1}{2}$	BIFURCATE TOGETHER	bound*
f	A	$A+\frac{1}{2}$	A	$A+\frac{1}{2}$		pace
g	A	B	$B+\frac{1}{2}$	A	$A = \frac{1}{2}$ period	
h	A	B	B	$A+\frac{1}{2}$	$B = \frac{1}{2}$ period	canter?
2 a	A	A	A	A		pronk
b	A	$A+\frac{1}{2}$	A	$A+\frac{1}{2}$		pace
c	A	A	$A+\frac{1}{2}$	$A+\frac{1}{2}$		bound*
d	A	$A+\frac{1}{2}$	$A+\frac{1}{2}$	A		trot
3 a	A	A	A	A		pronk
b	A	A	$A+\frac{1}{2}$	$A+\frac{1}{2}$		bound*
c	A	$A+\frac{1}{2}$	$A+\frac{3}{4}$	$A+\frac{1}{4}$		walk and amble
d	A	$A+\frac{1}{2}$	$A+\frac{1}{4}$	$A+\frac{3}{4}$	BIFURCATE	
e	A	$A+\frac{1}{2}$	$A+\frac{1}{2}$	A	TOGETHER	trot
f	A	$A+\frac{1}{2}$	A	$A+\frac{1}{2}$		pace
g	A	A	B	$B+\frac{1}{2}$	$A = \frac{1}{2}$ period	
h	A	$A+\frac{1}{2}$	B	B	$B = \frac{1}{2}$ period	
4 a	A	A	B	B		asymmetric bound*
b	A	$A+\frac{1}{2}$	B	$B+\frac{1}{2}$		
5 a	A	A	B	B		asymmetric bound*
b	A	$A+\frac{1}{2}$	B	B		
c	A	A	B	$B+\frac{1}{2}$		

*bound is close to transverse and rotary gallops

Our analysis does not force any commitment on whether a CPG is a self-sustaining system, or whether it is a forced system, responding to an incoming signal. Neither does it imply that a CPG must necessarily be truly 'central': its components might be distributed throughout the animal's body.   The patterns of symmetry-breaking are the same in all cases, provided only that the incoming signal is distributed identically to the component oscillators.

Paying attention to the symmetries of gaits may provide information on the likely structure of GPGs, and on how the connections to limbs should be arranged to produce the observed gaits.   Because varying parameters may produce different oscillation patterns in the same network layout, the same CPG may control a whole range of apparently very different gaits.

References

1   R.Abraham and J.E.Marsden, *Foundations of Mechanics* (2nd ed.), Benjamin/ Cummings, New York 1978.

2   R.Abraham and C.D.Shaw, *Dynamics: The Geometry of Behaviour* (4 vols.), Aerial Press, Santa Cruz 1983.

3   R.McN.Alexander, Terrestrial locomotion, in *Mechanics and Energetics of Animal Locomotion* (eds. R.McN.Alexander and G.Goldspink), Chapman nad Hall, London 1977, 168-203.

4   V.I.Arnold, Small denominators I: Mappings of the circumference onto itself, *Amer. Math. Soc. Transl.* Ser 2 **46** (1965) 213-284.

5   D.K.Arrowsmith and C.M.Place, *An Introduction to Dynamical Systems*, Cambridge University Press, Cambridge 1990.

6   P.Ashwin, Symmetric chaos in systems of three and four forced oscillators, *Nonlinearity*.

7   P.Ashwin, G.P.King, and J.W.Swift, Three identical oscillators with symmetric coupling, *Nonlinearity*.

8   P.Ashwin and J.W.Swift, The dynamics of *n* weakly coupled identical oscillators, preprint, Mathematics Institute, University of Warwick 1990.

9   C.Baesens, J.Guckenheimer, S.Kim, and R.S.MacKay, Three coupled oscillators: mode-locking, global bifurcations, and toroidal chaos, preprint, Mathematics Institute, University of Warwick 1990.

10   S.Castro, *Experiments in Nonlinear Dynamics: an Attractor of the Modified van der Pol oscillator*, Nonlinear Systems Laboratory Report, University of Warwick 1990.

11   J.J.Collins and I.N.Stewart, Coupled nonlinear oscillators and the symmetries of animal gaits, *preprint* **4/1990**, Mathematics Institute, University of Warwick, 1990.

12   P.Gambaryan, *How Mammals Run: Anatomical Adaptations*, Wiley, New York, 1974.

13   P.A.Getting, Comparative analysis of invertebrate central pattern generators, in *Neural Control of Rhythmic Movements in Vertebrates* (eds. A.H. Cohen, S. Rossignol, and S. Grillner), Wiley, New York 1988, 101-127.

14   L.Glass and M.Mackey, *From Clocks to Chaos: the Rhythms of Life*, Princeton University Press, Princeton 1988.

15   M.Golubitsky and I.N. Stewart, Hopf bifurcation in the presence of symmetry, *Arch. Rational Mech. Anal.* **87** (1985) 107-165.

16   M.Golubitsky and I.N. Stewart, Hopf bifurcation with dihedral group symmetry: coupled nonlinear oscillators, in *Multiparameter Bifurcation Theory* (eds. M. Golubitsky and J. Guckenheimer), Contemporary Math. **56**, Amer. Math. Soc., Providence 1986, 131-173.

17   M.Golubitsky and I.N. Stewart and D.G. Schaeffer, *Singularities and Groups in Bifurcation Theory*, vol. II, Springer, New York 1988.

18   S.Grillner, Locomotion in vertebrates: central mechanisms and reflex interaction, *Physiol. Rev.***55** (1975) 247-304.

19   J.Guckenheimer and P.Holmes, *Nonlinear Oscillations, Dynamical Systems, and Bifurcations of Vector Fields*, Springer-Verlag, New York 1986.

20   C.Hayashi, *Nonlinear Oscillations in Physical Systems*, McGraw-Hill, New York 1964.

21   M.Hildebrand, Symmetrical gaits of horses, *Science* **150** (1965) 701-708.

22   E.A.Jackson, *Perspectives of Nonlinear Dynamics* (2 vols), Cambridge University Press, Cambridge 1990.

23   N.Kopell, Toward a theory of modelling central pattern generators, in *Neural Control of Rhythmic Movements in Vertebrates* (eds. A.H. Cohen, S. Rossignol and S. Grillner), Wiley, New York 1988, 369-413.

24   R.Rand, A.H.Cohen and P.J.Holmes, Systems of coupled oscillators as models of central pattern generators, in *Neural Control of Rhythmic Movements in Vertebrates* (eds. A.H.Cohen, S.Rossignol and S.Grillner), Wiley, New York 1988, 333-367.

25   J.C-Roux, R.H.Simoyi, and H.L.Swinney, Observation of a strange attractor, *Physica* 8D (1983) 257-266.

26   G.Schöner, W.Y.Yiang, and J.A.S.Kelso, A synergetic theory of quadrupedal gaits and gait transitions, *J. Theor. Biol.* **142** (1990) 359-391.

27   H.G.Schuster, *Deterministic Chaos: an Introduction*, Physik-Verlag, Weinheim 1984.

28   A.I.Selverston, Are central pattern generators understandable?, *Behavioral and Brain Sci.* 3 (1980) 535-571.

29   J.W. Swift, Hopf bifurcation with the symmetry of the square, *Nonlinearity* 1 (1988) 333-377.

30   J.M.T.Thompson and H.B.Stewart, *Nonlinear Dynamics and Chaos*, Wiley, New York 1988.

Identical Oscillator Networks with Symmetry

Peter Ashwin*
Mathematics Institute,
Warwick University,
Coventry CV4 7AL
United Kingdom.

Abstract

This is a review of some theoretical and experimental results concerning networks of identical oscillators with S_n symmetry under a hypothesis of weak coupling. Such networks have a global or mean field coupling so that the network is symmetric under any interchange permuting the oscillators. The phase space is highly structured by the weak coupling and symmetry, and allows some quite unusual dynamical transitions to take place. Upon forcing the system, there is experimental evidence of symmetric bifurcations of chaotic attractors.

1   Introduction: coupled oscillators

When answering questions about the behaviour of a physical system, it is always necessary to balance the complexity of a mathematical model with its usefulness in providing answers. Many researchers have turned their attention to the study of coupled oscillators as a heuristic model for some physical system. For example, the biological sciences provide systems of such complexity that fundamental modelling is very difficult, but coupled oscillator models have been used with great success (cell morphogenesis [1, 2], small intestine motion [3] or animal gaits [4]). Winfree and others [5, 6] consider populations of biological oscillators and show synchronisation to be a very general phenomena. Studies have been made of arrays with frequency gradients, and phase-locking into plateaux [5, 7, 3, 8, 9] has been noted. Detailed studies of small numbers of oscillators [10, 11] have shown that a wide variety of behaviour is possible using simple assumptions, and that coupled oscillators can provide a useful context in which to see many types of dynamical periodic or quasiperiodic behaviour.

In the theory of oscillations, electronic oscillators have played a big part in inspiring work from the time of van der Pol [12], and later [13, 14]. Electronic

---

*Supported by a British Gas Research Scholarship. Present address: FB Mathematik, Universität Marburg, D-3550 Marburg, FRG

circuits give easily controllable and measurable low-noise systems that demostrate e.g. period doubling and chaotic behaviour in physical systems such as the Ruelle-Takens route to chaos by using three weakly coupled oscillators [15].

More recently, interest in Josephson junction arrays [16, 17, 18] has led to studies of large populations of oscillators which are *identical*, an important requirement being that the oscillators synchronise in-phase. (cf. [19] which is a study of synchronisation in three identical microwave oscillators.) In this article we review some analytical, electronic and numerical studies of small numbers of mutually interacting oscillators that have appeared in [20, 21, 22] and concentrate on three oscillator systems.

1.1 Coupling strength

In order to be able to call individual cells of a network oscillators, it is necessary that they have at least a propensity to oscillate. We shall do this by defining a network that has a parameter governing coupling strength allowing decoupling of the cells. In this limit, the *uncoupled* state, the cells will perform oscillations independently and at any instant a system of n oscillators can be described by a point on an n-dimensional torus $\mathbf{T}^n$, as each oscillator can be specified by its phase. For weak enough coupling strengths, this torus persists and provides substantial simplifications to the dynamics. This contrasts with the approach of [23, 24] who define cells at a point of Hopf bifurcation, and allow strong coupling of the same order of strength as the internal dynamics.

2 Identical oscillators

In the following section, $\mathbf{T}$ refers to the circle group $\mathbf{R}/2\pi\mathbf{Z}$, and Θ is a point in the lift of $\mathbf{T}^n$ with components θ_i.

2.1 Near Hopf bifurcation

In [25], Golubitsky and Stewart perform analysis of Hopf bifurcation with Γ some Lie group of symmetries. They use Lyapunov-Schmidt reduction to reduce to an irreducible representation of Γ, and find a normal form with $\Gamma \times \mathbf{T}$ symmetry. They then focus on the particular cases $\mathbf{Z}_n$ (and $\mathbf{D}_n$) of a ring of n cells with (resp. without) a preferred direction, calculating the structure of generic smooth germs at bifurcation, to discover which bifurcation diagrams may occur generically. This analysis is taken considerably further by Swift [26] who concentrates on the case of four oscillators with $\mathbf{D}_4$ symmetry.

2.2 The weak coupling limit

A limit cycle is a most fundamental concept of dynamical behaviour in a system. If we have a spatially extended system composed of a clearly definable set of units with a propensity to oscillate, and coupling between them, then we might describe this to be a network of coupled oscillators. In order to make a mathematical definition, we define the system to be a perturbation of an uncoupled system, either using a natural 'strength of coupling' parameter or artificially with some homotopy parameter. In many applications, we are concerned with groups of identical or near identical oscillators arranged in some network with possibly some symmetry:

An **oscillator** is a system governed by an ODE of the form:

$$\dot{x} = f(x, \lambda) \tag{1}$$

where $x \in X = \mathbf{R}^m$ for some m and $\lambda \in \mathbf{R}$ is a parameter, such that the equation has an attracting limit cycle smoothly varying with λ.

A **network of identical weakly coupled oscillators** is a dynamical system governed by an ODE of the form:

$$\dot{\mathbf{x}} = \mathbf{f}(\mathbf{x}, \lambda, \epsilon) \tag{2}$$

where:

- $\mathbf{x} \in \mathbf{X} \equiv X^n$ is the phase variable, $X = \mathbf{R}^m$ for some m

- $\lambda \in \mathbf{R}$ is the bifurcation parameter

- $\epsilon \in \mathbf{R}$ is the coupling strength parameter

- $\mathbf{f}$ is smooth in x, λ and ϵ.

- $\mathbf{f}(\mathbf{x}, \lambda, 0) = (f(x_1, \lambda), \cdots, f(x_n, \lambda))$ for some oscillator $\dot{x} = f(x, \lambda)$ on X which has a stable limit cycle $\{\gamma(t) : t \in T\}$ with period normalised to 2π for convenience.

2.3 Averaging on the invariant torus

By using normal hyperbolicity [27], we can for $\epsilon \ll 1$ reduce a dynamical system defined by equation 2 to one on an n-torus, and when this exists the system is **weakly coupled** [9, 10, 11]. The uncoupled system has an n-torus with $n + 1$ Floquet exponents equal to zero, and normal hyperbolicity causes this torus to persist for small enough perturbations. The system also inherits an approximate **T** symmetry which is made rigorous by averaging the system to separate the timescales of fast oscillations and slow relative changes in phase. The question of what happens when the coupling is big enough that the invariant torus breaks down is addressed in section 2.7.

The weak coupling assumption allows us to give the equations a normal form:

$$\dot{\theta}_i = 1 + \epsilon F_i(\Theta, \lambda) \tag{3}$$

where $\Theta = \{\theta_1, \ldots, \theta_n\}$, and $\mathbf{F}$ is equivariant under the action of the circle group:

$$F_i(\Theta) = F_i(\Theta + \lambda(1, \cdots, 1)).$$

It is usually very difficult to analytically derive an expression for $\mathbf{F}$, and most authors start with a model for $\mathbf{F}$ based on some physical motivation. The reduction will work even for strongly nonlinear oscillations, although the magnitude of the coupling must be small compared to the attraction of the limit cycle. Equivalently, it is possible to derive a global Poincaré return map on the two-torus $\mathbf{T}^{n-1}$ perpendicular to the $(1, \ldots 1)$ direction in $\mathbf{T}^n$. When the coupling is weak, the Poincaré map is near-identity, and it can be approximated by an ordinary differential equation. This near identity Poincaré map can be approximated by an ODE with ϵ^p-closeness for times of order ϵ^{-1} [28]. This time limit means it is necessary to be careful about drawing conclusions of dynamical equivalence. However, invariant subspaces forced by the symmetry are truly invariant, and hyperbolic fixed points are accurately portrayed for small enough ϵ.

2.4   Symmetries of the network: isotropy and the phase space

Even for oscillators which are identical in the limit of no coupling, this symmetry is broken upon addition of generic coupling. Any symmetry that occurs in a network can be included by defining the group of permutations $\Gamma < \mathbf{S}_n$ that leave the system unchanged upon permuting the set of n-oscillators labelled $\{1, \ldots, n\}$. As examples, oscillators in rings with or without directional coupling have symmetry groups $\mathbf{Z}_n$ or $\mathbf{D}_n$ respectively; for the second case, detailed work has been done by Ermentrout [29], and both have been investigated near Hopf bifurcation by Golubitsky and Stewart [23].

Networks with total permutational symmetry ($\mathbf{S}_n$) have been studied by many authors [30, 16, 17, 18] motivated by the dynamics of arrays of Josephson junctions. It is possible to envisage applications to networks with a variety of symmetries, not just indistinguishable oscillator networks. (A network of *indistinguishable* oscillators has a transitive permutation group of symmetries; thus by a suitable re-labelling of oscillators it is possible to permute any two oscillators in the network.) An example of a network with a distinguishable oscillator is a ring with a central oscillator; this is an n-oscillator network with $\mathbf{Z}_{n-1}$ symmetry.

We shall concentrate on $\mathbf{S}_n$ symmetric networks for small n. For these, it is possible to show [22] that the conjugacy classes of isotropy subgroups [31] for the $\mathbf{S}_n \times \mathbf{T}$ action on $\mathbf{T}^n$ are in 1-1 correspondance with all distinct ways of writing $n = m(k_1 + k_2 + \cdots + k_l)$, where l, m and k_j are all integers and $k_1 \geq k_2 \geq \cdots \geq k_l \geq 1$. The members of conjugacy classes are of the form:

$$\Sigma_{\mathbf{k},m} \equiv (\mathbf{S}_{k_1} \times \cdots \times \mathbf{S}_{k_l})^m \otimes \mathbf{Z}_m$$

where $n = mk$ and $k = k_1 + k_2 + \cdots + k_l$ and the dimension (in $\mathbf{T}^n$) of $fix(\Sigma_{\mathbf{k},m})$ is l.

Any fixed point space of a group in $\Sigma_{k,m}$ induces a partition of the oscillators into m blocks, each containing k oscillators. The system is unchanged under a time shift of $1/m$th of a period, coupled with a cyclic permutation of the blocks. (This is the $\mathbf{Z}_m$ symmetry.) Each block is partitioned into sets of phase-locked oscillators of sizes k_i and thus has $\mathbf{S}_{k_1} \times \cdots \mathbf{S}_{k_l}$ symmetry.

In particular we are assured the existence of periodic orbits with total $\mathbf{S}_n$ symmetry, called in-phase oscillations, and with $\mathbf{Z}_n$ symmetry, called rotating waves (also called *splay states* [18], *ponies on a merry-go-round* [30] or from [19] *n-phase oscillations*). This isotropy structure can be further elucidated by defining the **canonical invariant region** [22] (the largest path-connected dynamically invariant region in the lift of $\mathbf{T}^n$):

$$C = \{\Theta : \theta_1 < \theta_2 < \cdots < \theta_n < \theta_1 + 2\pi\}.$$

The whole canonical invariant region C is in fact an n-simplex in $\mathbf{T}^n$, but with only $\mathbf{Z}_n$ symmetry rather than $\mathbf{S}_n$. At the geometrical centre of each of these invariant regions there is a rotating wave.

2.4.1 Internal symmetries of the oscillators

If the oscillators have internal symmetries in addition to those of the network, sometimes these can interact to provide non-geneic behaviour; for example subspaces foliated with periodic orbits [24, 26]. This may happen for $\mathbf{D}_{4m}$ networks with $\mathbf{Z}_2$ symmetric waveforms, or $\mathbf{S}_{km}$ networks with $\mathbf{Z}_k$ symmetries in the phase plane of the oscillators [22], under an assumption of *pairwise coupling*. Thus, in order to see generic bifurcations, it is best to try coupling oscillators without symmetry, or at least with a symmetry of order coprime to the number of oscillators in the group.

2.5 Three oscillators with $\mathbf{S}_3$ symmetry

For a group of three identical weakly coupled indistinguishable oscillators, having used normal hyperbolicity, we have an ODE on $\mathbf{T}^3$:

$$\dot{\theta}_1 = 1 + \epsilon f(\theta_1; \theta_2, \theta_3) \tag{4}$$

(plus cyclic permutation of the labels for the other two oscillators), where $f : \mathbf{T} \to \mathbf{R}$. Because of the symmetry f satisfies:

$$f(\theta_1; \theta_2, \theta_3) = f(\theta_1; \theta_3, \theta_2)$$

Define

$$\psi = \frac{1}{3}(\theta_1 + \theta_2 + \theta_3) \in \mathbf{R}$$

and

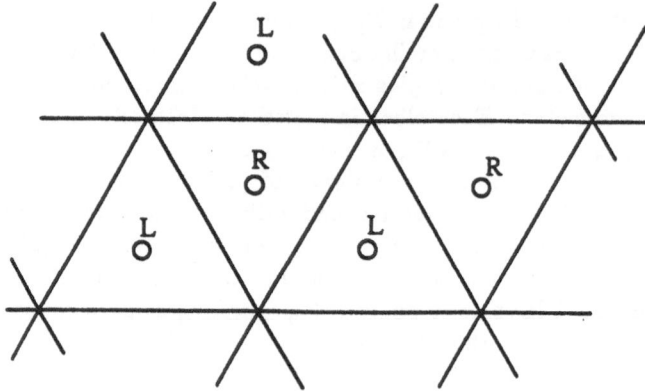

Figure 1: The two torus T^2 shown as a lift in the complex ϕ plane. The triangles are images of the canonical invariant region, while two together cover the torus. The in-phase oscillation is at the intersection of three lines. The two rotating wave states corresponding to the two possible orders 1,2,3 and 1,3,2 are at the centres of the equilateral triangles. Fixed points on the lines are the 2-in-phase oscillations.

$$\phi = \omega\theta_1 + \omega^2\theta_2 + \theta_3 \in \mathbf{C}$$

(So that ψ is invariant and ϕ is equivariant under the permutations in Γ.) The general ODE on T^3 can then be written in these coordinates to give

$$\dot{\psi} = 1 + \epsilon g(\psi, \phi)$$

$$\dot{\phi} = \epsilon h(\psi, \phi)$$

where g (resp. h) is invariant (resp. equivariant) under the action of S_n on (ψ, ϕ).

The canonical fundamental region in this case is an equilateral triangle. On the torus T^2 this corresponds to one half of a generating parallelogram, and the vector field is generated by knowing it in one sixth $(1/|S_3|)$ of the generating parallelogram and looking at the orbit of the vector field under the action of S_3. There is a web of dynamically invariant lines, the *2-in-phase* lines (fixed point spaces with S_2 isotropy in which $\theta_i = \theta_j$ for $i \neq j$). These features are pictured in figure 1.

2.6 Bifurcations on the invariant torus with S_3 symmetry

2.6.1 The $S_3 THB$

At the in phase solution, there must be two equal real eigenvalues. When they cross through zero there is locally an S_3 ($= D_3$) transcritical bifurcation, [31] with normal form:

$$\dot{\phi} = \mu\phi + \bar{\phi}^2$$

with $\mu \in \mathbf{R}$ and bifurcation at $\mu = 0$. If we assume that the in-phase solution goes unstable with no more than the minimum number of fixed points on the '2-in-phase' invariant manifolds, there is an interesting global bifurcation shown in figure 2 using the topology of the torus. This the S_3 transcritical/homoclinic bifurcation or S_3THB. At the bifurcation point there is a connection in the lattice between adjacent in-phase solutions, provided there are no other fixed points of the ODE on the invariant line. In the torus, this is a homoclinic orbit, while in the lattice it is a heteroclinic loop. The homoclinic orbit is either attracting or repelling, unless there is a constant of the motion. If it is attracting, then there is a stable limit cycle on the side of the bifurcation where the in-phase solution is unstable. (This situation is shown in figure 2.) If, on the other hand, the heteroclinic loop is repelling, then an unstable periodic orbit co-exists with the stable in-phase solution. (This is obtained by time reversal of figure 2.)

The stability of this homoclinic connection is unusual, as unlike homoclinic connections without, or with Z_2 symmetry, the stability is governed by the flow along the invariant sides of the triangle rather than being dominated by the flow near the fixed point. In the case without symmetry, the limit cycle is stable if the ratio of the eigenvalues at the fixed point are greater than 1 in modulus. Here, the in-phase solution has a double eigenvalue of zero at bifurcation, and so the limit cycle is very weakly attracting or repelling as it passes near the fixed point.

At the bifurcation point, the stability is dominated by the flow along the edges of the triangle; if we consider a trajectory approaching a stable S_3THB by reinjecting to a neighbourhood of the fixed point after going through a linear contraction $x \mapsto \mu x$ with $0 < \mu < 1$, as the linear part is zero the approach to the heteroclinic will go like $x_0 \mu^n$ after n approaches to the fixed point. It is possible to show that the local dynamics gives the transit time past the fixed point as $K\mu^{n/3}$. This means the period around the whole loop (three approaches to the fixed point) will go like $K\mu^n$, as we would expect from one approach to a hyperbolic fixed point, the generic case without symmetry.

2.6.2 A global bifurcation scenario: transition from in-phase to rotating stability

The simplest generic transition from the rotating wave to the in-phase wave (experimentally observed in the next section) is:

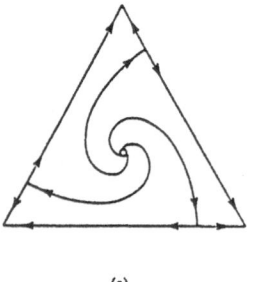  

(a)   (b)   (c)

Figure 2: The S_3 transcritical/homoclinic bifurcation (S_3THB): (a)$\mu < 0$, (b)$\mu = 0$, (c)$\mu > 0$. There is an S_3 transcritical bifurcation in (b)Simultaneous with this is a triangular heteroclinic connection between the three in-phase solutions. On the torus this is a homoclinic connection. When the heteroclinic loop at $\mu = 0$ is stable, as shown here, there must be a stable limit cycle very close to the triangle for $0 < \mu \ll 1$.

- The two rotating waves are the only stable solutions.

- There is a supercritical Hopf bifurcation of each of the two rotating waves.

- Only the two limit cycles are stable, one in each of the invariant regions of the ϕ torus. The limit cycles have Z_3 symmetry and correspond to a torus in the original system.

- The limit cycles grow until they are destroyed at an S_3 transcritical homoclinic bifurcation which stabilises the in-phase solution.

- The in-phase oscillation is now stable.

This sequence is one of the simplest consistent with the S_3 symmetry; figure 3(a) shows a bifurcation diagram for this scenario.

In a more general case, but still assuming there are no fixed points other than the minimum necessary (i.e. six in the torus), there will be a global branch of tori starting from the Hopf bifurcation and ending at the S_3THB, as shown in figure 3(b). This branch can fold back on itself several times, creating a sequence of saddle-node bifurcations of tori.

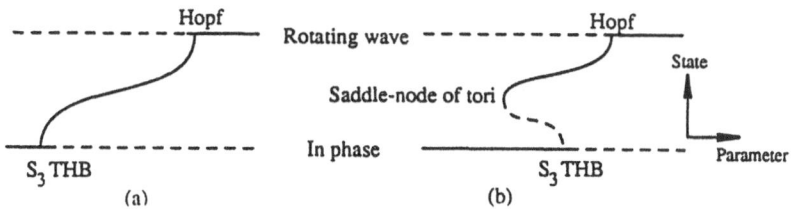

(a) (b)

Figure 3: Some typical bifurcation diagrams showing a branch of tori joining the rotating and in-phase solutions. (a) is a direct connection, whereas (b) folds over at a saddle node of tori bifurcation.

2.7 Break-up of the invariant torus

Upon increasing the strength of coupling or forcing a system undergoing stable periodic oscillations with an oscillator at an incommensurate frequency, the invariant torus will break up as described in [32]. If the periodic orbit has lower symmetry than the system, then at some point it is natural to expect we should get a crisis [33] where conjugate attractors of low symmetry collide to create an attractor of full symmetry. Several authors [34, 35] have investigated systems with Z_2 symmetry that is first broken by steady-state bifurcations and then recovered by a crisis in the attractors. Chossat and Golubitsky [36] hypothesised the existence of generic symmetry increasing bifurcations of attractors, where a group orbit of attractors can collide to form an attractor of higher symmetry, and work along this line is reviewed in [37].

3 Experiments

3.1 An electronic van der Pol oscillator

The oscillator consists of a parallel inductor-capacitor-resistor (LCR) network, the circuit being shown in figure 4.

3.2 Electronic simulation of an S_3 network of oscillators

Three van der Pol oscillators were coupled through a low-pass filter made of a star configuration of resistors earthed through a capacitor at the central node, as shown in figure 5. This provides a network with approximate S_3 symmetry

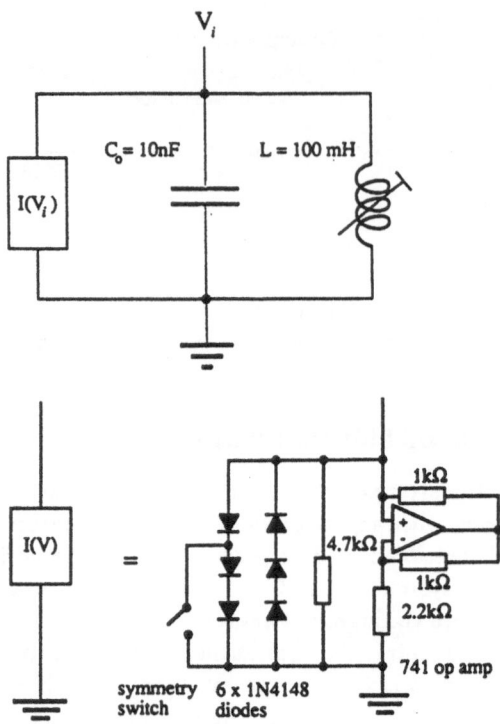

Figure 4: (a) The circuit for one van der Pol oscillator. (b) The nonlinear $I(V)$ characteristic is synthesised by a negative impedance converter and a set of diodes to define the saturation. The 'symmetry switch' allows the oscillators to produce symmetric or asymmetric waveforms.

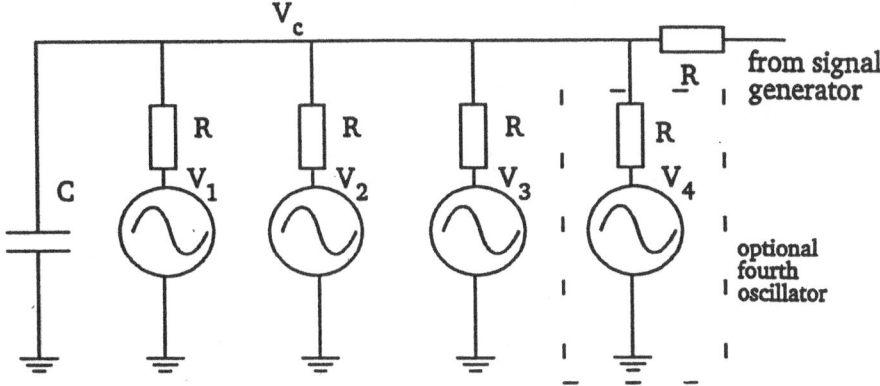

Figure 5: The coupling is achieved through a simple low-pass filter, and the value of C is varied as a bifurcation parameter. The network is symmetric under any interchange of the oscillators. An extra input is provided to allow forcing of the system.

and a control to switch between in-phase and rotating wave stability. The coupled circuit is a seven degree of freedom system, due to two dimensions in each oscillator and the extra dimension of the voltage at the centre of the coupling network. An input was provided as shown, in order to be able to force the system externally from a signal generator whilst preserving the symmetry of the circuit, but initially this was left unconnected.

The output from the three oscillators was visualised on an oscilloscope using the projection:

$$V_x = \frac{1}{2}(-V_1 - V_2 + 2V_3),\ V_y = \frac{\sqrt{3}}{2}(V_1 - V_2)$$

This is the same projection used to go from $(\theta_1, \theta_2, \theta_3)$ to the complex angle ϕ, noting that

$$V_x + iV_y = \omega V_1 + \omega^2 V_2 + V_3$$

This linear combination is an equivariant projection of the three individual voltages onto the plane. Cyclic permutation of the three oscillators gives a rotation by $2\pi/3$ in the $V_x - V_y$ plane, and a permutation of two oscillators gives a reflection.

The equations of motion for our system of oscillators may be derived using Kirchoff's current law after rescaling time so that $dx/dt = \omega_0 \dot{x}$, where $\omega_0 = (LC_0)^{-1/2}$ is the natural frequency, we obtain the non-dimensional equations:

$$\ddot{V_i} + \Gamma(V_i)\dot{V_i} + V_i\ =\ \epsilon(\dot{V_c} - \dot{V_i}) \tag{5}$$

$$\dot{V_c} = \delta \sum_{i=1}^{3}(V_i - V_c) + A\cos\nu t$$

with a dimensionless dissipation function $\Gamma(V) = \sqrt{(L/C_0)}dI/dV$. For the classical van der Pol model $\Gamma(V) = a + bV^2$. There are two parameters in this setting of the problem:

$$\epsilon = \frac{1}{\omega_0 RC_0} \quad \delta = \frac{1}{\omega_0 RC}$$

For the oscillator components given in figure 4, and $R = 10$ k, the numerical values of the dimensionless parameters are $\epsilon = 0.316$ and $\delta = 0.316C_0/C$.

3.2.1 Results without forcing

With $A = 0$, results for $R = 10$ k, and C treated as the bifurcation parameter are shown in this section. The time series (not shown) are quite far from sinusoidal. Figure 6 shows a sequence of transitions obtained when the individual oscillators have asymmetric waveforms. For large C, the two rotating waves are the only stable solutions. As C is decreased, each rotating wave undergoes a supercritical Hopf bifurcation to a stable torus (corresponding to a limit cycle in the ϕ-plane). This torus grows until it comes very close to the in-phase solution. Becuase of imperfections in the symmetry of the system, the S_3THB shown in figure 3(a) is replaced by a sequence of bifurcations but nonetheless, the in-phase solution is uniquely stable at low C. With symmetric waveforms, the scenario was closer to that shown in figure 3(b).

3.2.2 Results with forcing

The network in figure 5 was forced from marginal stability of the rotating wave by a sinusoidal signal with a frequency incommensurate to that of the oscillations. For small amplitude forcing, the system gives two-frequency quasiperiodic motion built up from the frequencies of the oscillators and those of the forcing signal. This corresponds to an attracting 2-torus for the dynamics, and is observable as a limit cycle in the Poincaré section. The same network was also extended to four asymmetrical oscillators using a D_4 equivariant projection of an S_4 symmetric system.

The signal projected onto (V_x, V_y) and sampled at the zero crossing of the forcing signal to give a stroboscopic picture. Transients were allowed to die away before starting to sample.

Figures 7 and 8 show three and four oscillator systems forced at two different frequencies. For different frequencies, the details of the transitions change, but in general there is a mixture of subharmonic and Hopf bifurctions to give chaotic attractors of low symmetry. In particular the results show initial breakdown of the 2-torus via saddle-node of limit cycle, Hopf and 2-torus doubling bifurcations [32]. The resulting chaotic attractors can collide to create ones with higher symmetry, in a way that [36, 38] describe.

33

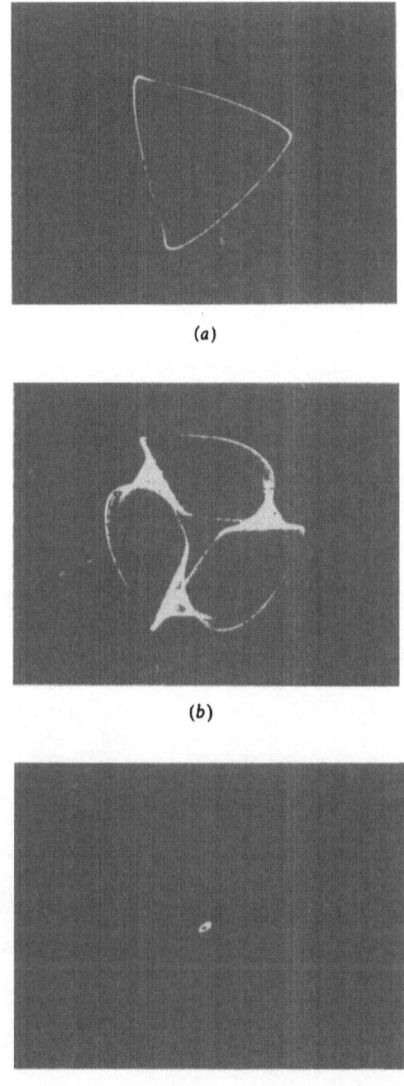

(a)

(b)

(c)

Figure 6: The transition from rotating wave to in-phase stability. (a) For large
C, the rotating wave is stable. (b) The torus bifurcating from the rotating wave.
(c) The in-phase oscillation for low C.

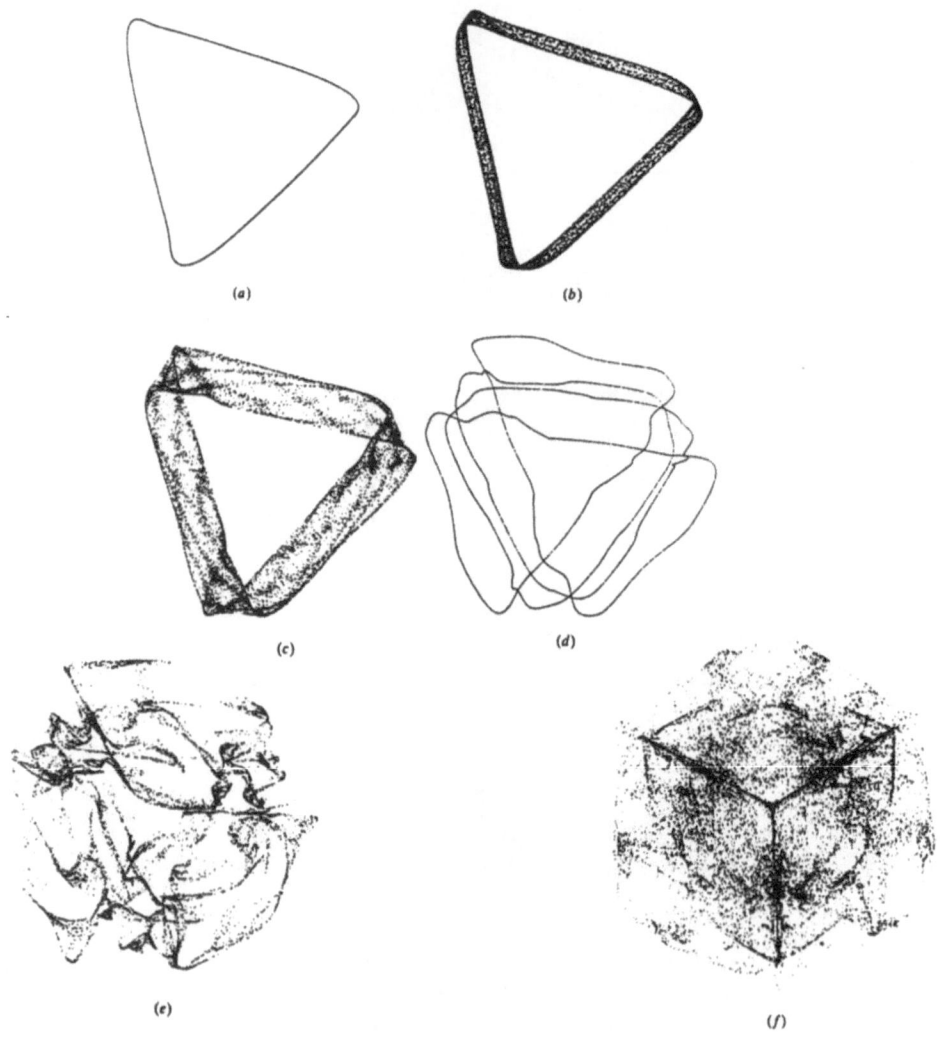

Figure 7:
(a) to (f) show a sequence of plots for samples of the output from three identical oscillators coupled with S_3 symmetry, sinusoidally forced and sampled at the forcing frequency as the forcing amplitude increases from zero. Note the symmetry increasing bifurcation (e)-(f).

35

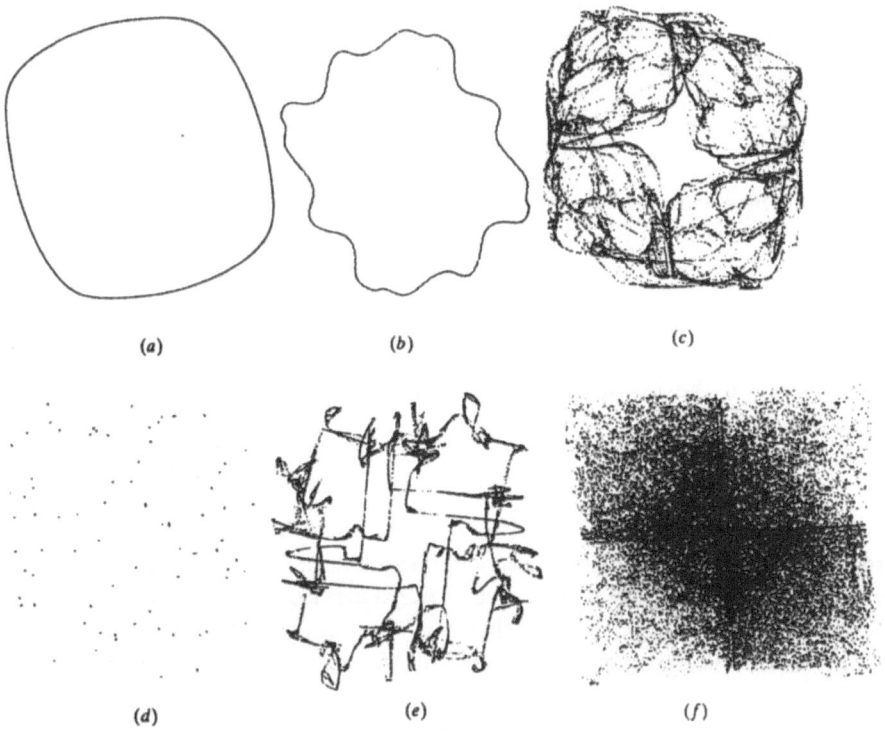

Figure 8:
A bifurcation sequence for four forced oscillators starting at a Z_4 rotating wave. The figures show a D_4 projection sampled at the forcing frequency, and the amplitude of forcing increases from (a)-(f).

Phenomena	Numerical δ	Experimental δ
Asymm. Hopf	0.19	0.17
Asymm. THB	0.32	0.31~ 0.46
Symm. Hopf	0.10	0.11
Symm. Saddle node	0.13	0.12

Table 1: A comparison between the locations of the bifurcations for the numerical and electronic experiments. Note that the experimental location of the THB is approximate due to symmetry imperfections in the circuit.

3.3 Numerical simulation

Measurements of the I-V characteristic of the nonlinearity of the van der Pol oscillator gave the model:

$$I(V) = -0.475V + 1.75 \times 10^{-5} \sinh(6.03V) \tag{6}$$

where I is measured in mA and V is measured in Volts. The model was fitted simply by noting that the steepness of the exponentials meant that at $V = 0$, the linear term is the slope of the curve. After removing this linear part, the two factors governing the exponential behaviour were found by fitting to two of the $I(V)$ observations. This gave models for Γ of the form in the asymmetric case:

$$\begin{aligned} \Gamma(V) &= -0.828 + 1.67 \times 10^{-4} \exp(6.03V) \\ &+ 5.01 \times 10^{-4} \exp(-18.1V) \end{aligned} \tag{7}$$

and in the symmetric case:

$$\Gamma(V) = -0.828 + 5.39 \times 10^{-5} \cosh(6.03V). \tag{8}$$

The model equations and real parameter values were used in the KAOS interactive dynamical systems package of Guckenheimer and Kim to investigated the asymmetric and symmetric oscillator scenarios, and good agreement was found both qualitatively and quantitatively, allowing for the fact that the symmetry of the numerically simulated oscillators was much greater than that for the electronic simulations.

The value of $\epsilon = 0.316$ was fixed in all the experiments. In the asymmetric case, there was a Hopf bifurcation at $\delta = 0.19$ giving rise to a stable 2-torus which grew in amplitude until its longest period went to infinity and the system approached the in-phase solution at $\delta = 0.32$. In the symmetric case, The Hopf bifurcation at $\delta = 0.10$ gave rise to a branch of tori which lost stability without the period tending to infinity, at some distance from in-phase solution, at $\delta = 0.13$. Table 1 shows that these results compare favourably with those from the electronic experiments.

Figures 9 shows the numerical results projected in the same way as that used for the previous pictures.

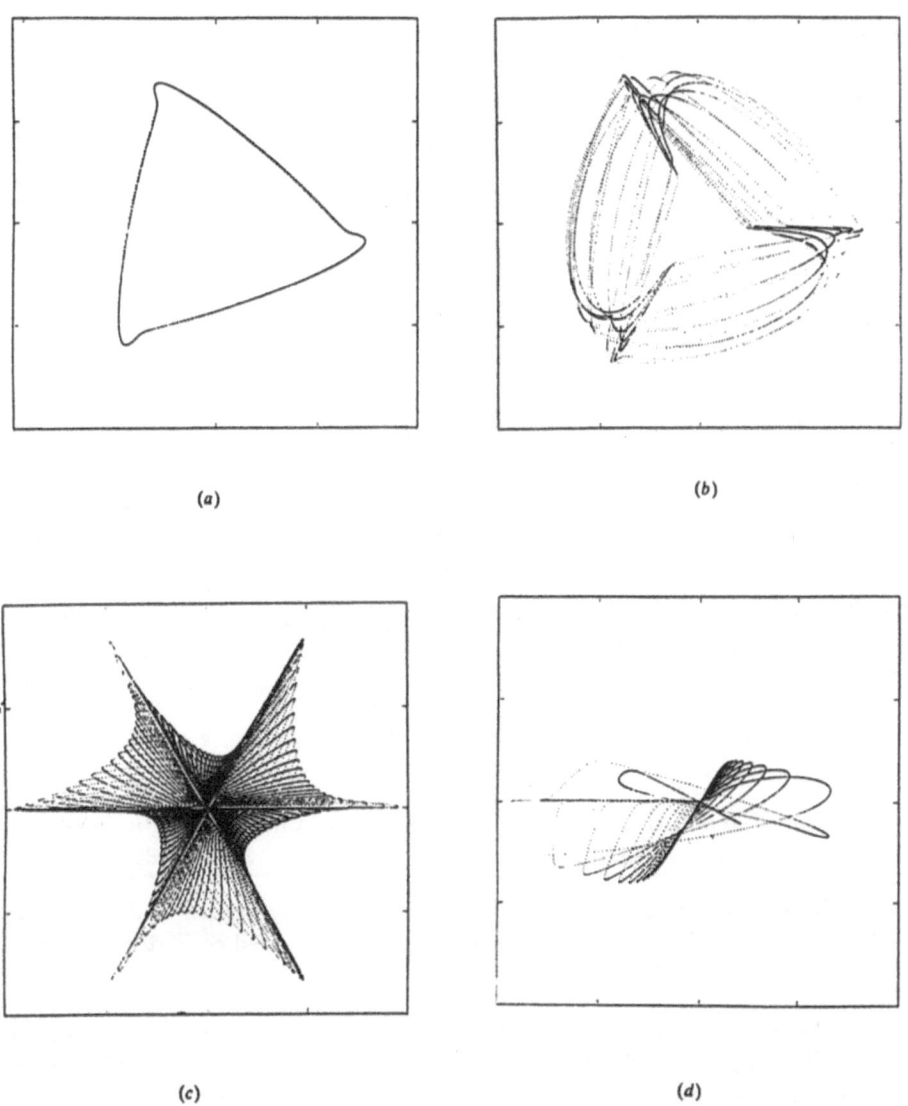

(a)

(b)

(c)

(d)

Figure 9: Numerical simulations with asymmetric waveform oscillators. (a) The rotating wave, $\delta = .106$ (b) After Hopf bifurcation, $\delta = .2$ (c) Near THB, $\delta = 0.31$ (d) Transient decaying to in-phase solution. Note that it approaches along a 2-in-phase direction. The boxes surrounding the pictures are $4V$ square.

References

[1] A.M. Turing. The chemical basis of morphogenesis. *Phil. Trans. Roy. Soc. Lond. B*, 237:37–72, 1952.

[2] H.G. Othmer and L.E. Scriven. Instability and dynamic pattern formation in cellular networks. *J. Theor. Biology*, 32:507–537, 1971.

[3] D. Linkens. Nonlinear circuit mode analysis. *IEEE proc*, 130A:69–87, 1983.

[4] J.J. Collins and I.N. Stewart. Coupled nonlinear oscillators and the symmetries of animal gaits. Preprint, University of Warwick, Mathematics Institute, 1990.

[5] A.T. Winfree. Biological rhythms and the behaviour of populations of coupled oscillators. *Theoretical biology*, 16:15–42, 1967.

[6] R.E. Mirollo and S.H. Strogatz. Synchronisation of pulse-coupled biological oscillators. *SIAM J. App. Math.*, 50:1645–1662, 1990.

[7] J. Grasman and M.J.W. Jansen. Mutually synchronised relaxation oscillators as prototypes of oscillating systems in biology. *J. Math. Biol.*, 7:171–197, 1988.

[8] G.B. Ermentrout and N. Kopell. Frequency plateaus in a chain of weakly coupled oscillators. *SIAM J. Math. Anal.*, 15:215–237, 1984.

[9] N. Kopell and G.B. Ermentrout. Symmetry and phase locking in chains of weakly coupled oscillators. *Comm. Pure App. Math.*, 39:623–660, 1986.

[10] D.G. Aronson, E.J. Doedel, and H.G. Othmer. An analytical and numerical study of the bifurcations in a system of linearly coupled oscillators. *Physica D*, 25:20–104, 1987.

[11] C. Baesens, J. Guckenheimer, S. Kim, and R.S. MacKay. Three coupled oscillators: mode-locking, global bifurcations and toroidal chaos. *Physica D*, 24:387–475, 1991.

[12] B. van der Pol. On relaxation oscillations. *Phil. Mag.*, 7-2:978–992, 1926.

[13] A.A. Andronov, A.A. Vitt, and S.E. Chaikin. *Theory of oscillators*. Pergamon, Oxford, 1966.

[14] C. Hayashi. *Nonlinear oscillations in physical systems*. McGraw Hill, 1964. Reprinted 1985 by Princeton University Press.

[15] P.S. Linsay and A.W. Cumming. Three-frequency quasiperiodicity, phase locking and the onset of chaos. *Physica D*, 40:196–217, 1990.

[16] P. Hadley, M.R. Beasley, and K. Wiesenfeld. Phase locking of Josephson junction series arrays. *Phys Rev B*, 38:8712–8719, 1988.

[17] P. Hadley. *Dynamics of Josephson junction arrays*. PhD thesis, Dept of Applied Physics, Stanford University, 1989.

[18] K. Wiesenfeld and P. Hadley. Attractor crowding in oscillator arrays. Preprint, Dept. of Physics, Georgia Tech., 1988.

[19] I. Ohta and T. Kaneko. Parallel running system of three oscillators coupled through a six-port magic junction. *IEEE Trans MTT*, 37:1699–1707, 1989.

[20] P. Ashwin, G.P King, and J.W. Swift. Three identical oscillators with symmetric coupling. *Nonlinearity*, 3:585–603, 1990.

[21] P. Ashwin. Symmetric chaos in systems of three and four coupled oscillators. *Nonlinearity*, 3:604–618, 1990.

[22] P. Ashwin and J.W. Swift. The dynamics of n identical oscillators with symmetric coupling. *Journal of Nonlinear Science*, 1992. (to appear).

[23] M. Golubitsky and I.N. Stewart. Hopf bifurcation with dihedral group symmetry. In *Multiparameter bifurcation theory*, volume 56 of *Contemporary Maths*, pages 131–173. AMS, Providence, RI, 1985.

[24] J.C. Alexander and G. Auchmuty. Global bifurcations of phase-locked oscillators. *Arch. Rat. Mech. Anal.*, 93:253–270, 1986.

[25] M. Golubitsky and I.N. Stewart. Hopf bifurcation in the presence of symmetry. *Arch. Rat. Mech. Anal.*, 87:107–165, 1985.

[26] J.W. Swift. Hopf bifurcation with the symmetry of the square. *Nonlinearity*, 1:333–377, 1988.

[27] M. Hirsch, C. Pugh, and M. Shub. *Invariant Manifolds*, volume 583 of *LNM*. Springer, Berlin, 1977.

[28] J.A. Sanders and F. Verhulst. *Averaging methods in nonlinear dynamical systems*. App. Math. Sci. 59. Springer, New York, 1985.

[29] G.B. Ermentrout. The behavior of rings of coupled oscillators. *J. Math. Biol.*, 23:55–74, 1985.

[30] D.G. Aronson, M. Golubitsky, and M. Krupa. Coupled arrays of Josephson junctions and bifurcations of maps with S(n) symmetry. Preprint, Dept of Mathematics, University of Minnesota, 1989.

[31] M. Golubitsky, I.N. Stewart, and D. Schaeffer. *Groups and singularities in bifurcation theory volume 2*, volume 69 of *App. Math. Sci.* Springer, New York, 1988.

[32] R.S. MacKay and C. Tresser. Transition to topological chaos for circle maps. *Physica D*, 19:206–237, 1986.

[33] C. Grebogi, E. Ott, and J.A Yorke. Crises, sudden changes in chaotic attractors and transient chaos. *Physica D*, 7:181–200, 1983.

[34] M. Kitano, T. Yabuzaki, and T. Ogawa. Symmetry-recovering crises of chaos in polarization-related optical bistability. *Phys. Rev. A*, 29:1288–1296, 1984.

[35] K.G. Szabó and T. Tél. On the symmetry-breaking bifurcation of chaotic attractors. *J. Stat. Phys.*, 54:925–948, 1989.

[36] P. Chossat and M. Golubitsky. Symmetry-increasing bifurcation of chaotic attractors. *Physica D*, 32:423–436, 1988.

[37] G.P. King and I.N. Stewart. Symmetric chaos. In Ames W. and Rogers C., editors, *Nonlinear Equations in the Applied Sciences*. Academic press, 1991.

[38] M. Krupa and R.M. Roberts. Symmetry breaking in equivariant circle maps. Preprint, Maths Institute, University of Warwick, 1991.

Bifurcating Neurones

A.V.Holden, J.Hyde, M.A.Muhamad and H.G.Zhang

Department of Physiology and Centre for Nonlinear Studies
University of Leeds
Leeds, LS2 9JT, UK

Abstract

The rich behaviour of neurones - resting states, regular repetitive discharges, patterned discharges and bursts, and irregular discharges are identified with qualitatively different solutions of differential systems that are separated in parameter space by bifurcations. Bifurcation curves are constructed for some neural models and methods of identifying bifurcation points from experimental recordings are discussed.

1. Neurones as dynamical systems.

Nervous systems are made up of large numbers (10^5-10^8 for invertebrates, up to 10^{13} for vertebrates) of interconnected neurones. Each neurone is a spatially extensive, complicated, system. However, what is biologically significant is not the spatio-temporal pattern of activity in a neurone but when this influences other cells. This is the temporal pattern of activity at its output synapses, the pre-synaptic membrane potential. This signal may be considered as the output from a dynamical system [1] and the nervous system as a network of interacting dynamical systems [2].

1.1. Isopotentiality.

The biophysical analysis of the membrane conductance systems of neurones requires preparations where dV/dx, the derivative of membrane potential V with distance, is zero: under these isopotential conditions the ionic currents flowing through the membrane charge or discharge the membrane capacitance, and so the preparation may be represented by a system of ordinary differential equations (ODEs):

$$C_m dV/dt = - I_{ion} = -f(V). \qquad (1)$$

A major success of membrane biophysics has been the quantitative description of the f(V) in terms of voltage dependent conductances and their kinetics: such excitation equations are nonlinear, and often of high order [3,4]. Biophysical excitation equations are only available for a few excitable membranes, where a detailed voltage clamp analysis of an isopotential preparation is possible. Even in these cases the description is continually evolving, as new conductances are characterised, and so an approach is to consider either generalised excitation equations, whose parameters may be adjusted to fit particular preparations, or simple polynomial caricatures, such as the FitzHugh-Nagumo [5] equations. An alternative approach is to consider the types of behaviour exhibited by neurones as corresponding to different qualitative behaviours of unspecified systems of ODEs [6,7].

1.2. Activity of isolated neurones.

Early electrophysiological studies on excitation were on isolated nerve trunks, whose axons have a stable resting state and which generate a propagating, all-or-none action potential in response to supra-threshold stimulus. Chronic recordings from neurones *in situ* in unanaesthetised animals, or from isolated neuronal somata show that activity - a regular or patterned discharge of action potentials - is common, and that a stable resting potential is just one possible type of behaviour, others being small amplitude oscillations, a regular repetitive discharge of action potentials (beating), or repetitive patterned bursts of action potentials. In invertebrates, where it is possible to identify the same neurone (in terms of size, shape, colour, connectivity and electrophysiology) in different individuals, an identified neurone has a characteristic discharge pattern (*e.g.* the parabolic burster R12 of *Aplysia* [8]) that can be changed by a variety of experimental conditions. These experimental procedures can be

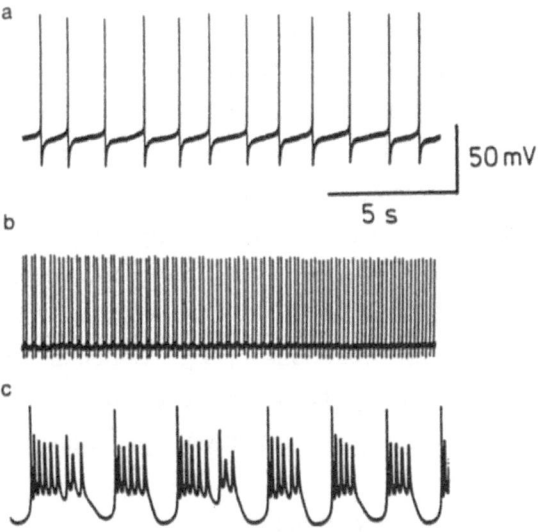

Figure 1. (a) beating, (b) patterned and (c) bursting activity recorded from different identified molluscan neurones.

considered as changing parameters in the system determining the behaviour, and the changes in behaviour as resulting frombifurcations. Different identified neurones in the same species, and even neurones and other excitable cells in different species all have the same sorts of membrane conductances, and so it is possible to encompass the baroque diversity of the electrophysiology of excitable cells by a common excitation equation, with different cells corresponding to different sets of parameter values for this excitation equation. Different cell types can be considered to be separated by bifurcation curves in parameter space, and the difference between a neurone with a resting membrane potential and an endogenously active neurone is not the presence of some pacemaker mechanism, but lies in the parameter dependence of the stability of the equilibrium state. To proceed with this approach we need to understand the bifurcations that occur in excitation equations, and to be able to identify bifurcation points when we have a time series outputs from cells, which are described by an unspecified excitation equation.

2. Axonal membrane.

Axons are long thin processes specialised for the rapid transmission of signals by propagating action potentials;

invertebrates achieve rapid conduction velocities by increasing the axonal diameter in giant axons, which are usually part of escape systems. These giant axons are unusual, in that their normal mode of behaviour is the resting state, and the occasional propagation of a solitary action potential to trigger an escape response; other axons normally sustain repetitive propagating wave trains of action potentials. The excitable membrane of the giant axon of the squid was the first excitable membrane to be voltage clamped [9,10,11] and the Hodgkin-Huxley equations [12], although for a peculiar excitable membrane, have been subject to extensive analysis [13,14,15] as if they were typical excitation equations.

2.1. The Hodgkin-Huxley equations

The squid axonal membrane only has three conductance systems, g_{Na}, g_K and g_L in parallel with a capacitance C_m. The membrane current I_m is given by

$$I_m = I_{Na} + I_K + I_L + C_m \, dV/dt$$

$$(2)$$

$$= g_{Na} (V - V_{Na}) + g_K (V - V_K) + g_L (V - V_L) + C_m \, dV/dt$$

In voltage clamp experiments, the membrane potential is held constant by external electronic circuitry, and the current required to hold the membrane potential constant is equal and opposite to the membrane ionic current, in this case $I_{Na} + I_K + I_L$. By varying the composition of the bathing fluid the ionic currents I_{Na}, I_K and I_L may be separated, from which $g_{Na}(V,t)$, $g_K(V,t)$ and g_L are calculated. The H-H equations provide an empirical description of the voltage-dependent kinetics of the conductances of the squid giant axon. The conductance of one of the conductance systems is treated as a fraction of the maximal observed conductance of that system: for the sodium system

$$g_{Na} = (m^3 h) \, \bar{g}_{Na}$$

where $\bar{g}_{Na}$ is the maximum conductance and $(m^3 h)$, $0 < (m^3 h) < 1$, is a function of voltage and time. A simple interpretation of this formalism is that the sodium conductance mechanism is a collection of discrete channels, any one of which is either open (conducting) or closed (non-conducting) with a single channel conductance which may be voltage-dependent. (m^3h) would represent the fraction of channels which were in their conducting state.

The sodium conductance responds to a maintained change in V by a transient increase in conductance: m represents a fast, activation process and h a slower inactivation process. $0 < m, h < 1$, and these activation and inactivation processes may be considered as gating mechanisms which control the channel - the channel is conducting if all gates are in the state which permits conduction. The exponent of 3 for the m system is required to fit the rising phase of the sodium conductance transient.

For the potassium system, $g_K = n^4 \bar{g}_K$; there is no inactivation process. The exponent of 4 is necessary to reproduce the delayed, sigmoid rise in g_K, and can be interpreted as the number of gates/channel. The leakage conductance g_L is a constant.

The voltage- and time-dependence of the m, n and h variables are separated by considering the first-order rate coefficients α_m, β_m, α_n, β_n, α_h, β_h:

$$dm/dt = \alpha_m(1 - m) - \beta_m m, \quad dn/dt = \alpha_n(1 - n) - \beta_n n \qquad (3)$$

$$dh/dt = \alpha_h(1 - h) - \beta_h h$$

$$\alpha_m = 0.1(25-E)/(\exp\{(25 - E)/10\} - 1),$$
$$\beta_m = 4 \exp(-E/18)$$
$$\alpha_h = 0.07 \exp(E/20),$$
$$\beta_h = 1/(\exp\{(30 - E)/10\} + 1)$$
$$\alpha_n = 0.01(10 - E)/(\exp\{(10 - E)/10\} - 1),$$
$$\beta_n = 0.125 \exp(- E/80)$$

where $E = V - V_{resting}$. Hodgkin and Huxley used the convention that the resting membrane potential $V_{resting}$ was set as 0 and a depolarization was negative. In this paper the resting potential of the unmodified H-H equations is taken to be -60 mV and depolarization moves in the positive direction.

$$g_{Na} = m^3h\,\bar{g}_{Na}, \quad g_K = n^4\,\bar{g}_K. \qquad (4)$$

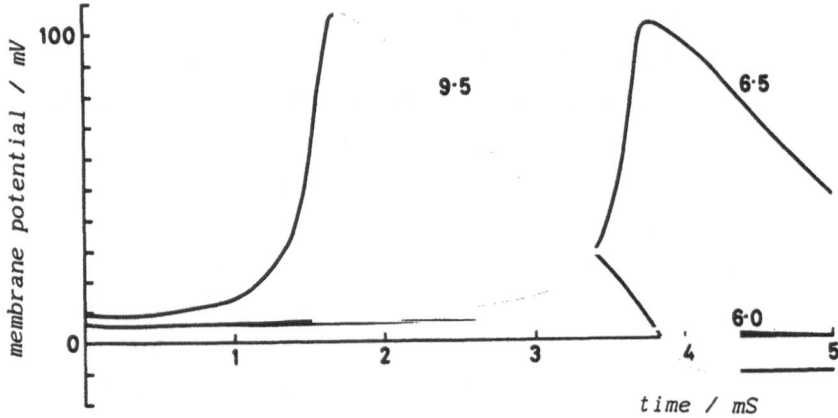

Figure 2. Numerical solutions of HH equations with standard parameter values for different initial conditions: the membrane potential returns to its stable equilibrium solution, via a small ampitude (subthreshold) or large amplitude excursion (all-or-none action potential).

Numerical solutions of the HH equations are shown in Figure 2. The HH equations (2-4) with standard parameter values have a globally attracting stable equilibrium point that correspond to the resting state of the membrane; as a parameter is changed this stable equilibrium can lose its stability at a bifurcation point.

2.2 Hopf bifurcation.

A Hopf bifurcation is from a stationary point to small amplitude periodic solutions and occurs when a single complex conjugate pair of eigenvalues cross the imaginary axis with a nonzero velocity, while the real part of the other eigenvalue remains negative. The small-amplitude periodic solutions that emerge at a Hopf bifurcation are stable at a supercritical bifurcation and are unstable at a subcritical bifurcation. In the Hodgkin-Huxley system Hopf bifurcations occur when the parameter I is changed at two points, I_1 and I_2 (see Figure 4 below).

To examine the solution for I near I_k k=1, 2, we introduce $\mu =$ I-I_k the bifurcation parameter. Let $\alpha(\mu) \pm i\omega(\mu)$ denote the complex conjugate pair of eigenvalues of the Jacobian matrix of the system

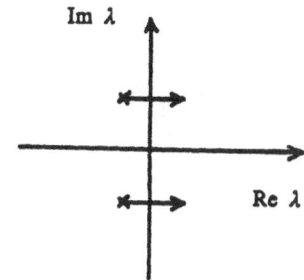

Im λ

Re λ

Figure 3. Hopf bifurcation: a single complex pair of eigenvalues crossing the imaginary axis with non-zero velocity.

such that $\alpha(0) = 0$ and let $\omega_0 = \omega(0) > 0$. If $\alpha'(0) \neq 0$, a theorem due to Hopf may be applied to demonstrate the existence of frequency of small amplitude solutions for small μ. The frequency f = f(ε) satisfies

$$\lim_{\varepsilon \to 0} f(\varepsilon) = \omega_0 / 2\pi$$

Hopf theory implies that $\mu(\varepsilon)$ and $f(\varepsilon)$ expand as

$$\mu(\varepsilon) = \mu_\varepsilon \varepsilon^2 + O(\varepsilon^4)$$

$$f(\varepsilon) = \frac{\omega_0}{2\pi}\left[1 + \tau_2 \varepsilon^2 + O(\varepsilon^4)\right]$$

For the unstable oscillations in the vicinity I_1

$$I - I_1 = \mu_2 \varepsilon^2 + O(\varepsilon^2)$$

and $\mu_2 < 0$.

The frequency of the small amplitude oscillations occur for values of I less than I_1 and since $\alpha'(0) < 0$, the oscillations are unstable. The amplitude and frequency of the stable oscillations obtained by numerical integrations may be compared with the equations

$$\varepsilon \approx \left[(I - I_1) / \mu_2\right]^{1/2}$$

$$f \approx \frac{\omega_0}{2\pi}\left[1 + \tau_2 \varepsilon^2\right]$$

Troy [16] used the results from his bifurcation analysis of the FitzHugh equation to guide investigations into the current-clamped H-H membrane equations. As $\tau_n >> \tau_m$ and $\tau_h >> \tau_m$ the fast V, m and slow n, h processes may be separated. For ε small and positive, the reduced system

$$I = C_m \, dV/dt + m^3 h \, \bar{g}_{Na}(V-V_{Na}) + n_\infty^4 \, \bar{g}_K(V-V_K) + g_L(V-V_L),$$

$$dm/dt = \gamma_m(V) \, (m_\infty(V) - m), \qquad\qquad\qquad (5)$$

$$dh/dt = \varepsilon \gamma_h(V) \, (h_\infty(V) - h),$$

where $\gamma = 1/\tau = \alpha + \beta$, has large amplitude periodic solutions. This reduced system retains the voltage-dependence of the sodium system, while leaving g_K at its resting value and emphasizes the importance of the transient sodium conductance. The restrictions used to obtain this result are quite general, and require:

(a) $V_K < V_L < V_{Na}$;

(b) the functions controlling the conductances are analytic and non-negative on [0,1], with positive differentials;

(c) γ_m, γ_h, γ_n and m_∞, h_∞ and n_∞ are analytic and positive, with appropriate asymptotes at $\pm \infty$ of 0 and 1;

(d) dm/dV is sufficiently large;

(e) $I = m^3 h \, \bar{g}_{Na}(V-V_{Na}) + n_\infty^4 \, \bar{g}_K(V-V_K) + g_L(V-V_L)$ has the characteristics of a cubic, that is there are no more than 3 distinct roots; and

(f) the V, m, h system has a unique steady-state solution.

Troy [17,18] uses Hopf bifurcation theory to identify critical values of I at which a bifurcation occurs. The n and h processes are treated as slow with respect to m and V. The existence of a bifurcation point is proved in the first paper, by considering the eigenvalues of the Jacobian matrix for the H-H membrane system. The second paper relates bifurcation theory results with numerical results. The bifurcation points, with bifurcation occurring to the left, are $I_0 = 6.2$, $I_1 = 9.8$ and $I_2 = 154 \ \mu A/cm^2$.

Hassard [19] uses powerful numerical methods described in [20] to locate the bifurcation points precisely, and finds a fourth

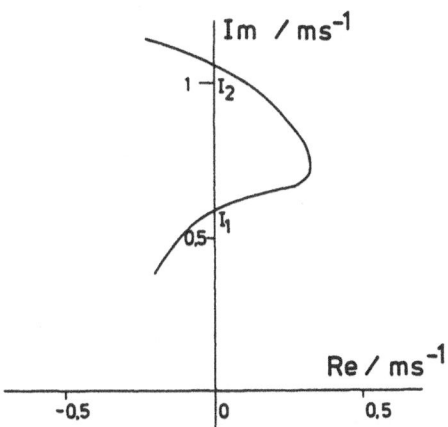

Figure 4. Illustration for locus of one of a complex conjugate pair of eigenvalues as the parameter I (applied current density) of Hodgkin-Huxley membrane equation is changed. As I is increased, loci of the complex conjugate pair of eigenvalues cross the imaginary axis when I = I₁ and then at I = I₂. Subcritical Hopf bifurcation occurs at I₁, supercritical Hopf bifurcation at I₂.

point I_3 at 7.8466 $\mu A/cm^2$, and suggests the existence of a fifth bifurcation between I_3 and I_1. Troy and Hassard's results are synthesized in Figure 4, where $||V||$ is plotted against I and the bifurcation points identified.

The differential system representing an excitable membrane contains a variety of parameters. These may include the ionic concentrations of the intra- and extracellular media, maximal ionic conductances or ionic channel densities, the concentration of biochemical and pharmacological agents, and applied currents. As an example, consider the applied current density I as a parameter:

$$dV/dt = -f(V) - I,$$

with $f(V)$ given by the HH equations

The equations are analytically intractable and solutions may be obtained only by numerical integration. However, the behaviour of solutions may be considered by examining the stationary points at which dV/dt, dm/dt, dn/dt, and dh/dt all equal zero, and so

$$f(V, m_\infty(V), n_\infty(V), h_\infty(V)) + I = 0 \qquad (6)$$

where $m_\infty(V) = \alpha_m(V)/\{\alpha_m(V) + \beta_m(V)\}$, and similarly for $n_\infty(V)$, $h_\infty(V)$. The eigenvalues of the Jacobian matrix

$$J = \begin{matrix} -\partial f/\partial V & -\partial f/\partial m & -\partial f/\partial n & -\partial f/\partial h \\ \gamma_m m'_\infty & -\gamma_m & 0 & 0 \\ \gamma_n n'_\infty & 0 & -\gamma_n & 0 \\ \gamma_h h'_\infty & 0 & 0 & -\gamma_h \end{matrix}$$

(7)

with $\gamma(V) = \alpha(V) + \beta(V)$ for m, n, and h, may be evaluated numerically at any stationary point. Under some conditions, the stationary points that are the solutions of (6) can form a curve that exhibits multiple equilibria: more than one value of V can satisfy (6) at a particular value of I. Multiple equilibria arise at bifurcation points where a single zero eigenvalue exits: at such a bifurcation point the stabiilty of the equilibrium solution changes.

If the eigenvalues of the Jacobian matrix (7), evaluated at stationary points, all have negative real parts, the stationary point is linearly stable, and the equilibrium solution corresponds to a resting membrane.

The eigenvalues λ of the Jacobian matrix (7) satisfy $(J - \lambda I) = 0$ and are the roots of the characteristic polynomial

$$\lambda^4 + C_1\lambda^3 + C_2\lambda^2 + C_3\lambda + C_4 = 0 \qquad (8)$$

where the coefficients $C_1,...,C_4$ are obtained from the Jacobian as
$$C_1 = \gamma_m + \partial f/\partial V + \gamma_n + \gamma_h,$$
$$C_2 = \gamma_m(\partial f/\partial V + m'_\infty \partial f/\partial m) + (\gamma_m + \partial f/\partial V)(\gamma_n + \gamma_h)$$
$$+ \gamma_n(n'_\infty \partial f/\partial n + \gamma_h) + \gamma_h h'_\infty \partial f/\partial h$$
$$C_3 = \gamma_m(\gamma_n + \gamma_h)(\partial f/\partial V + m'_\infty \partial f/\partial m) + \gamma_m(\gamma_n n'_\infty \partial f/\partial n$$
$$+ \gamma_h h'_\infty \partial f/\partial h) + \gamma_n\gamma_h(\gamma_m + \partial f/\partial V + n'_\infty \partial f/\partial n + h'_\infty \partial f/\partial h),$$
$$C_4 = \gamma_m\gamma_n\gamma_h(\partial f/\partial V + m'_\infty \partial f/\partial m + n'_\infty \partial f/\partial n + h'_\infty \partial f/\partial h). \quad (9)$$
Thus the eigenvalues can be evaluated at any stationary point by a straightforward but lengthy calculation.

The stability of stationary points of the HH equations may be changed by any parameter that influences the coefficients $C_1,...,C_4$. As such a parameter is changed monotonically, the eigenvalues will change until, at some critical value of the parameter, the stationary point loses its stability. This change from a stable to an unstable stationary point occurs as the real part of some of the eigenvalues cross the imaginary axis at a bifurcation point.

The points I_1, I_2 in Figure 4 occur close to 9.78, 154.53 $\mu A.cm^{-2}$. Close to such Hopf bifurcation points the period and amplitude of the small-amplitude periodic solutions that emerge may be evaluated from Hopf bifurcation theory. The agreement between these values and those obtained by numerical integration is good only close to the bifurcation point. This is illustrated for the amplitude of periodic solutions at current densities close to I_2 in Figure 5.

The standard Hodgkin-Huxley system has a single equilibrium solution at any value of the current density I. This equilibrium solution is stable for $I < I_1$, for $I_1 < I < I_2$ it is unstable, and it is stable for $I > I_2$. Between the two Hopf bifurcation points the only stable solutions are periodic. Near I_1 these are large-amplitude solutions that correspond to a repetitive discharge of action potentials; as I is increased further, the amplitude of these solutions decreases until they vanish at I_2, as in Figure 5.

However, for values of I just less than I_1, with $6.26 < I < I_1$ = 9.78 $\mu A.cm^{-2}$, both the equilibrium solution and large-amplitude periodic solutions are stable, and these stable solutions are connected by unstable solutions. In this narrow range of current densities, an appropriate small perturbation can flip activity between an equilibrium and periodic activity; the annihilation of periodic activity in this range has been demonstrated numerically and experimentally [21,22]. If, in the HH system I = 0 and $\bar{g}_K$ is reduced from its standard value of 36 $mS.cm^{-2}$ only one equilibrium solution exists at any $\bar{g}_K$, and this equilibrium solution exchanges stability at subcritical Hopf bifurcations close to 3.84 and 19.76 $mS.cm^{-2}$. A change in $\bar{g}_K$ may be interpreted as a change in membrane channel density, produced by a mismatch in the rates of channel production and breakdown, or a change in membrane area, and may be mimicked by blocking K^+-selective channels by tetraethylammonium (TEA) or 4-aminopyridine (4AP).

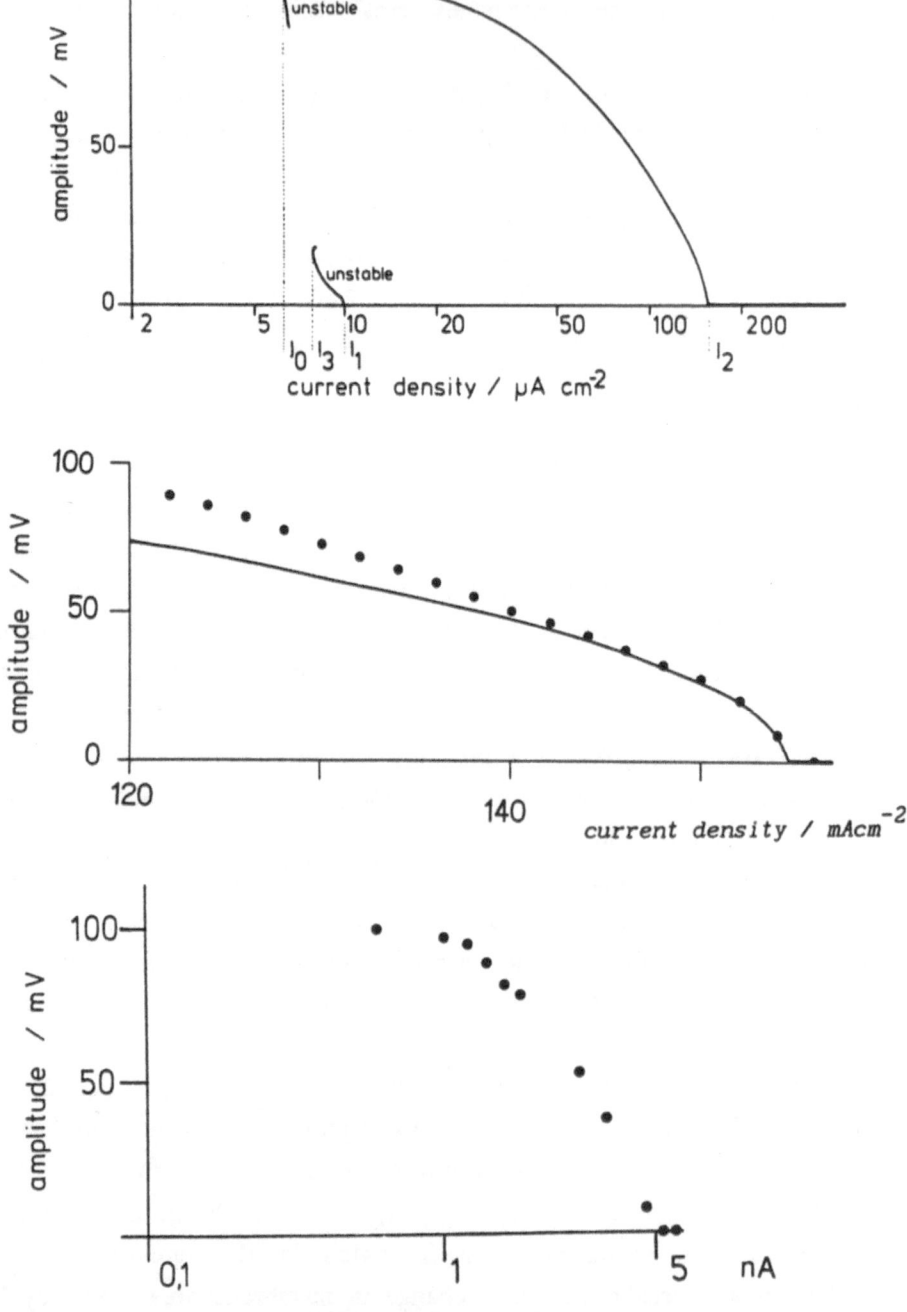

Figure 5. (a), (b). $\|V\|$ *against I for stable and unstable solutions of the H-H membrane equations. The bifurcation points I_o, I_1, I_2, I_3 are identified. (c) $\|V\|$ - I for a molluscan neurone - the observed action potentials are stable solutions. The curve drawn through the data points is obtained from bifurcation theory.*

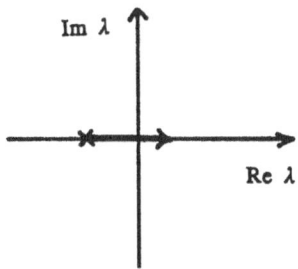

Figure 6. Fold (or saddle node) bifurcation .

2.3. Fold bifurcation and multiple equilibria.

A four variable system such as the HH equations has four eigenvalues; generally two real and a complex conjugate pair. If the real part of all the eigenvalues is less than zero then the equilibrium solution is stable, if, as a parameter is changed, one of the real eigenvalues passes through zero the equilibrium becomes unstable.

For the HH equations when either the applied current density I or the maximal K^+-conductance $\bar{g}_K$ of the Hodgkin-Huxley equations with I = 0 are varied alone as bifurcation parameters, the only bifurcations are from equilibria to small-amplitude periodic solutions at Hopf bifurcation points. However, when $(\bar{g}_K,I)$ is varied, multiple equilibria occur at low $\bar{g}_K$ and hyperpolarizing current densities - see Figure 7.

For $\bar{g}_K > 10.6$ mS.cm^{-2}, only one equilibrium solution exists at any I; for $\bar{g}_K$, 10.6 mS.cm^{-2} multiple equilibria appear in an area of the $\bar{g}_K$ - I plane that is the projection of a cusp. In this area one would expect to find in numerical integrations V at the upper (depolarized) or lower equilibria, not at the intermediate unstable equlibrium. At the boundaries of the area of multiple equilibria there is a single zero eigenvalue.

2.4 Bifurcation curves

If a parameter is varied, bifurcations will occur at a point; if two parameters are varied the bifurcations will occur on curves. These curves may be located by using path following algorithms [23] such as AUTO [24] or PATH [25], which, given a differential system

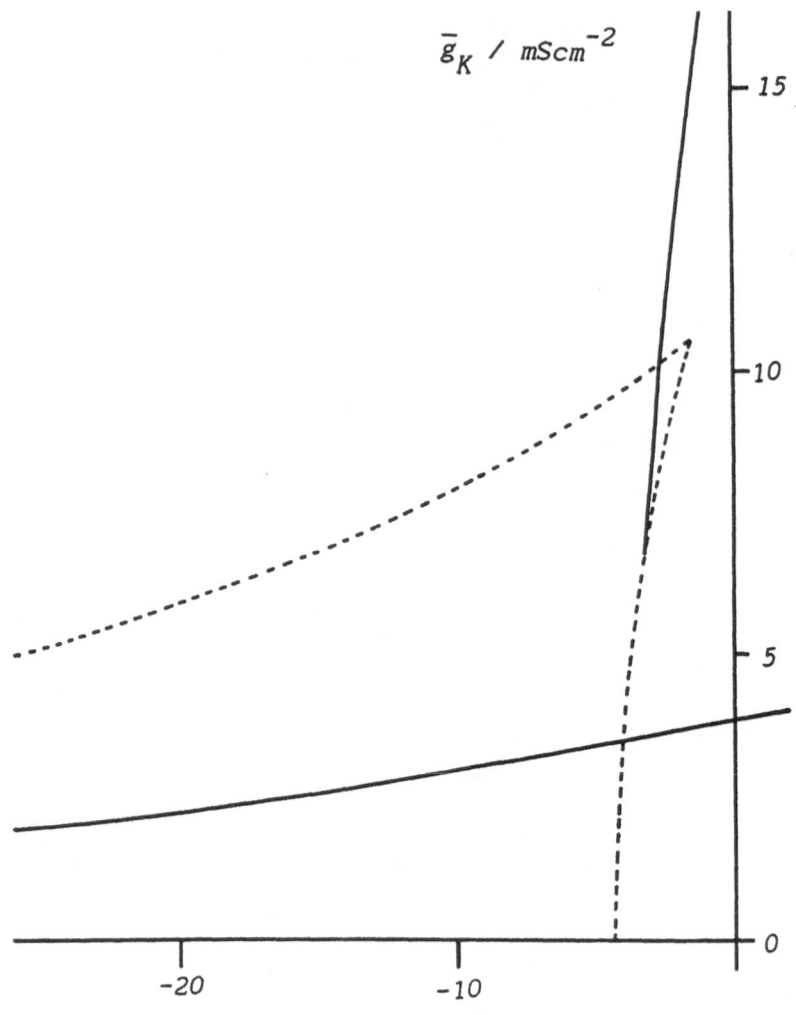

$$\bar{g}_K \ / \ mScm^{-2}$$

current density /μAcm^{-2}

Figure 7 (a) Area of $\bar{g}_K$ - I plane that supports multiple equilibria is enclosed by dashed lines, solid lines are Hopf bifurcations.

and an approximate solution, then finds the solution and evaluates its linear stability by finding all the eigenvalues of the Jacobian matrix of a stationary solution, and by evaluating the eigenvalues of the derivative of the Poincaré map. A path following algorithm tracks the curves where solutions lose their stability as parameters are changed.

When $(\bar{g}_K, I)$ is the bifurcation parameter, bifurcations will occur along curves in the $\bar{g}_K$ - I plane. The Hopf bifurcation curve

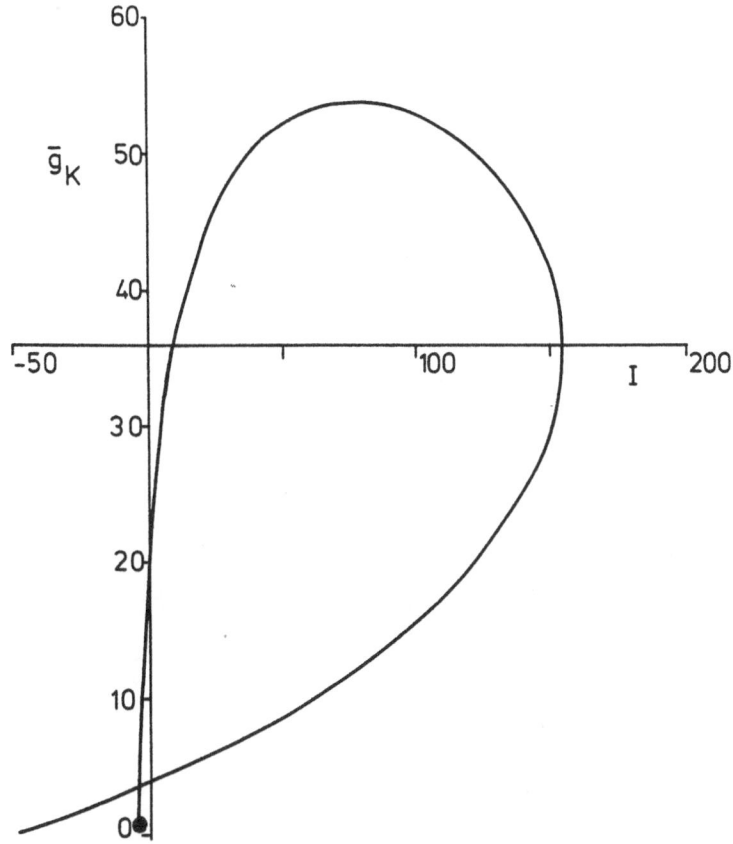

Figure 7 (b) Hopf bifurcation curves in the $\bar{g}_K$ - I plane.

in the $\bar{g}_K$ - I plane is shown in Figure 7b; the axes are the standard $\bar{g}_K$ (36 mS.cm^{-2}) and zero current. As one moves along the bifurcation curve, changes occur between sub- and supercritical bifurcations. The solid circle is the termination point of the curve of Hopf bifurcations. Both the real and imaginary parts of the single complex conjugate pair of eigenvalues become zero.

The effect of reducing $\bar{g}_K$ is to reduce the strength of the repolarising current; this can also be achieved by an increase in V_K, simulating an increase in extracellular $[K^+]$, as occurs in epileptic foci [26,27]. Bifurcation curves in the V_K-I plane are shown in figure 8. Singularity theory is used by Labouriau to relate these results to a Hopf-Takens bifurcation and a Bogdanov-Takens cusp in [28].

Bifurcation curves in the $\bar{g}_K$, $[Ca^{2+}]_o$ plane are presented in [29]. The result of these numerical studies is to predict that the exotic behaviour associated with multiple equilibria will occur

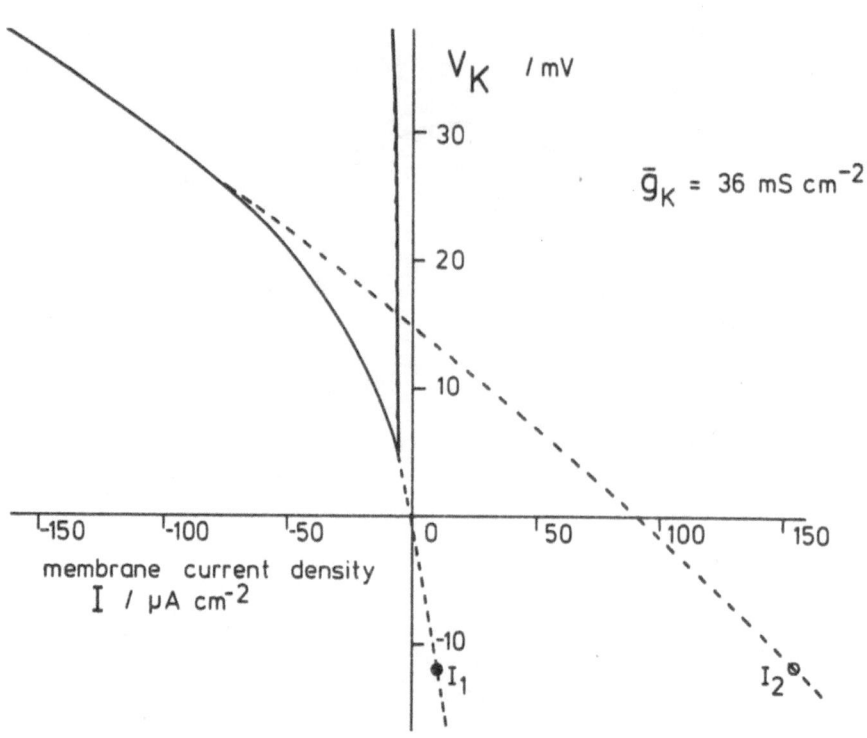

Figure 8. Bifurcation curves in V_K-I plane

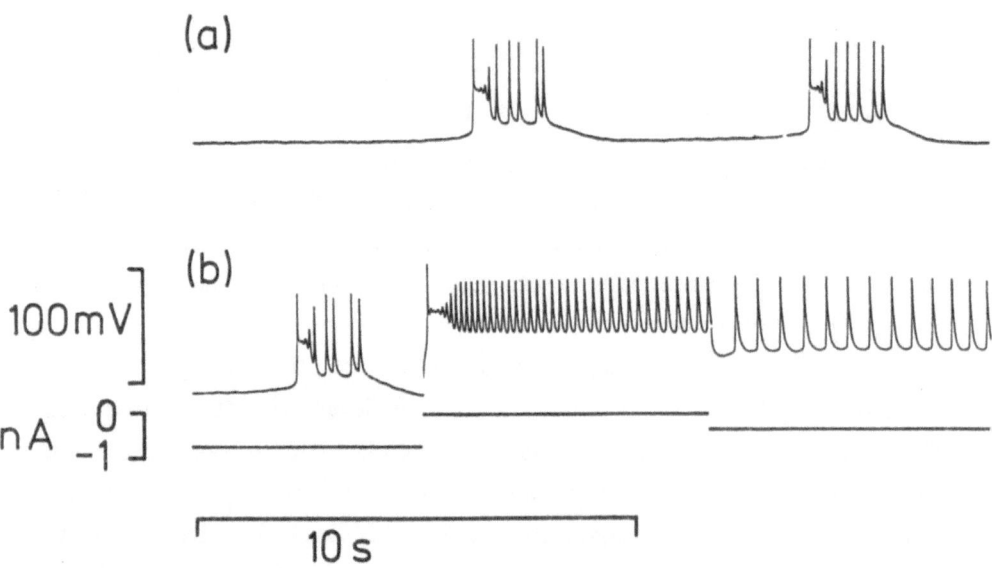

Figure 9. Behaviour of molluscan neurone in low $[Ca^{++}]$ and hyperpolarising current. The hyperpolarising current switches between regular oscillations and bursts

when either a reduced repolarising current (shown by K^+ conductance or increased V_K) or a reduced $[Ca^{2+}]_o$ exists, but only when either occurs in the presence of hyperpolarizing currents. Behaviour consistent with this has been described in molluscan neurones in low $[Ca^{++}]$; see figure 9 and [30].

2.5 Axonal behaviour.

In these numerical and bifurcation theoretical studies the onlybehaviours seen are either equilibria (stable or unstable) or simple periodicity (small or large amplitude; stable or unstable). The complicated, fourth order nonlinear H-H differential system shows the simple behaviour characteristic of a two-variable nonlinear oscillator, and in fact the Hodgkin-Huxley equations can readily be reduced to a two-variable system ($m(V) \cong m_\infty(V)$, as m is a fast variable, and $h \cong (1-n)$ empirically). For more complicated behaviour, such as patterned periodicity and chaos, another variable is necessary. This can be introduced by forcing, or by adding a charge dependent conductance.

3. Somatic membrane.

Vertebrate nerve cell somata are irregular in shape, and surrounded by a thin layer of extracellular fluid whose composition can change during activity. Thus a full voltage clamp analysis of their membrane conductances is not in general possible. However, some mammalian neurones have a simple geometry and may be voltage clamped - Kostyuk *et al.*[31] have developed a technique for voltage clamping vertebrate dorsal root ganglion cells under conditions where both the intracellular and extracellular fluid concentrations are under direct control. Tissue slice preparations allows two electrode voltage (but not space) clamping of hippocampal CA3 neurones. The giant somata of some molluscan ganglion cells are also readily voltage clamped. From experiments on these and other preparations it is possible to generalise that somatic membrane is more complicated than axonal membrane, in that it contains more conductance systems:

-fast and slow Na^+ selective conductance

-a slow, voltage dependent Ca^{++} selective conductance

-fast and slow K^+ selective conductances

-an outward current that is activated by hyperpolarisation

-Ca^{++}-activated, K^+ selective voltage independent current.

This richness in conductance systems allows a wide range of transient, oscillatory and chaotic behaviour. These behaviours - bifurcations between periodic solutions, and into chaos - can be obtained from 3-variable nonlinear systems of ODEs, and so it is not surprising that the full range of behaviour is retained by the

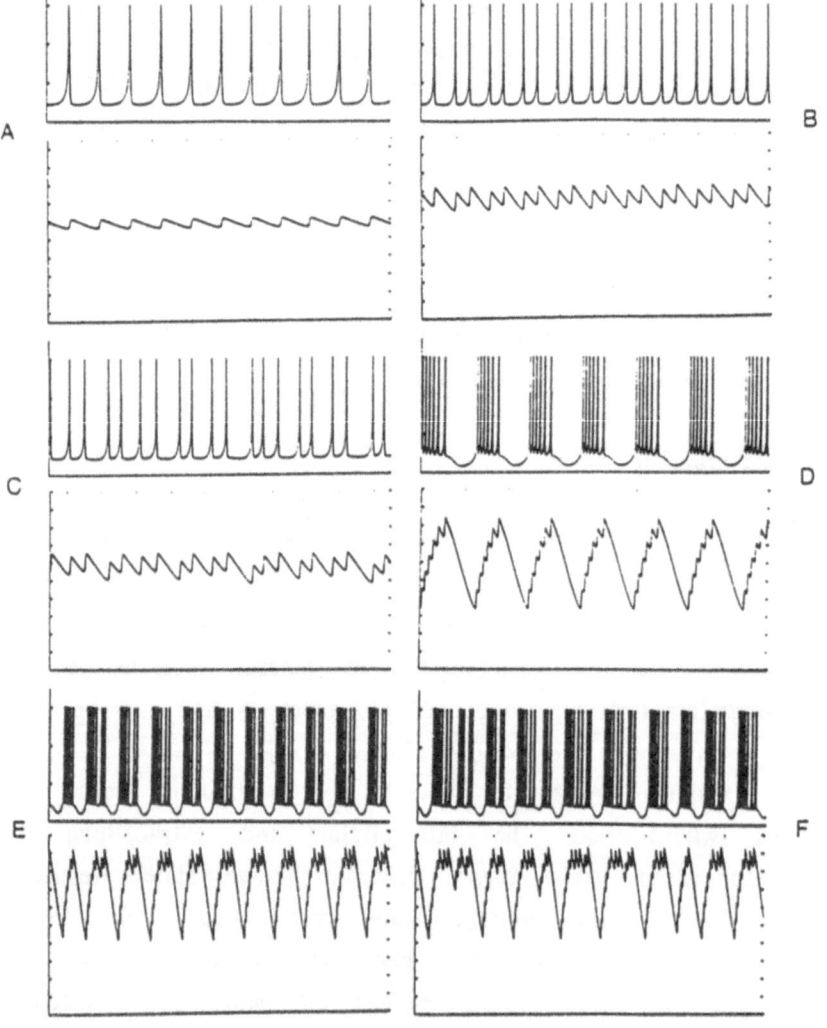

Figure 10 Numerical solutions of the Chay equations, with V(t) above C(t). Same voltage scale; different scales for C and time. (a) Simple, (b) period two, (c) period four, (d) six burst and (e) nine burst periodic solutions; (f) irregular (chaotic) solution. These solutions were obtained for different values of $g_{K,C}^*$

3-variable simplifications of the biophysical excitation equations such as the Chay equations.

3.1 The Chay equations.

Voltage clamp analyses of excitable membranes from different nerve and muscle cells have shown a wide range of conductance processes, all with H-H type kinetics, with α's and β's that are voltage dependent. In addition, there are charge dependent conductances, in that entry of Ca^{++} charge (current x time) activate a K-selective conductance. Chay [32] has simplified the variable model for pancreatic β cells [33] to give a 3 variable model for molluscan neuronal somatic membrane.

$$dV/dt = g_I^* m_\infty^3 h_\infty (V_I - V) + g_{K,V} n^4 (V_K - V)$$
$$+ g_{K,C}^* C/(1+C).(V_K-V) + g_L^* (V_L-V) + I$$
$$dC/dt = \rho\{m_\infty^3 h_\infty (V_C - V) - k_C C\} \tag{10}$$
$$dn/dt = (n_\infty - n)/\tau_n$$

Time t measured in ms is the independent variable. The dependent variables are V, the membrane potential in mV, C, the dimensionless calcium concentration, and n, a probability of activation. Numerical solutions of (10), with appropriate values for the parameters $g_{K.C}^*$ and I show simple periodic, doublet, triplet, bursting and chaotic discharges - see Figure 10.

3.2 Path following algorithms.

Biophysically plausible excitation equations are obtained by a detailed voltage clamp analysis of a single preparation, are usually complicated and of high order. Since they are obtained by a process of analysis in which parameters are estimated to obtain the best fit for experimental measurements of ionic currents, it is not altogether surprising that the excitation equations provide a good fit to voltage clamp measurements. The plausibility of the voltage clamp description of the electrical activity of the membrane as resulting from a series of gated membrane conductance systems is enhanced by the fact that the equations also provide a good description of the action potential, and transient and periodic trains of action potentials. However, this very success in providing an accurate description of the electrical behaviour of neuronal membrane raises problems in chosing the most appropriate excitation equation. Although there is no doubt that the large number of conductance systems exist, the (high order) biophysical excitation equations can be reduced to lower order systems (by separation of fast and slow processes, by combining different ionic conductances) to give a mathematically more tractable, low order system whose solutions still reproduce the behaviour of nerve membrane. The Chay equations are in fact an example of this, as they have been obtained by reducing a five variable system. To evaluate the behaviour of different excitation systems, and to gain insight into the the dynamical mechanisms of their behaviour, a method for comparing the behaviours of different excitation systems is needed. An approach is to compare their maintained behaviours in parameter space, and how they change as parameters are varied. This requires the construction of bifurcation diagrams by path following algorithms, as illustrated for the Hodgkin-Huxley equations in Figure 7, the Chay equations in Figure 11 and the Hindmarsh-Rose equations in Figure 14.

For the nonlinear system

$$g(V,\mu) = 0, \; g = \mathbb{R}^n \times \mathbb{R} \rightarrow \mathbb{R}^n$$

with $V \in \mathbb{R}^n$ and $\mu \in \mathbb{R}$, we need to calculate the solution set of $g(V,\mu) = 0$ near a point (V_0,μ_0) that satisfies $g(V_0,\mu_0) = 0$. The basis for continuation, or path-following, algorithms is the Implicit Function Theorem that ensures the existence of a smooth path of solutions $V = V(\mu)$ near (V_0,μ_0), with $V(\mu_0) = V_0$ as long as the n by n Jacobian matrix of g is nonsingular. To compute the neighbouring point corresponding to $\mu = \mu_0 + \Delta\mu$ for sufficiently small $\Delta\mu$ a predictor-corrector method is commonly used. AUTO [24] has been applied to the Chay equations [34] and a bifurcation diagram obtained using PATH [25] is given in Figure 11. In spite of the richness of behaviour seen in Figure 10 the bifurcation diagram is not very impressive; although there are some periodic to periodic bifurcations (period-doublings) many of the changes between bursts with different numbers of action potentials are not associated with bifurcations.

3.3 Period doubling bifurcations

A period doubling, or flip bifurcation, can readily be visualised in a state space projection of a periodic orbit as m orbits become $2m$ orbits. Typically, a stable limit cycle loses its stability, while another closed orbit is formed whose period is twice the period of the original cycle [35].

To understand period doubling in $\mathbb{R}^n$ space $(n \geq 2)$ it is advantageous to have a basic knowledge of Floquet theory [36]. A Floquet exponent is defined as $\sigma = \zeta + i\eta$ through the relation

$$\lambda(T) = \exp \sigma T$$

where T is the time period. $\lambda(T)$ is the *Floquet multiplier* and represents the unit circle. The periodic solution loses its stability when the complex conjugate pair of multipliers λ, $\bar{\lambda}$ escape from the unit circle at $\lambda = \pm 1$ (Figure 12)

3.4 Bursting and homoclinic orbits.

The numerical integration shown in Figure 10 show that, during repetitive activity (as during a burst) there is an increase in the

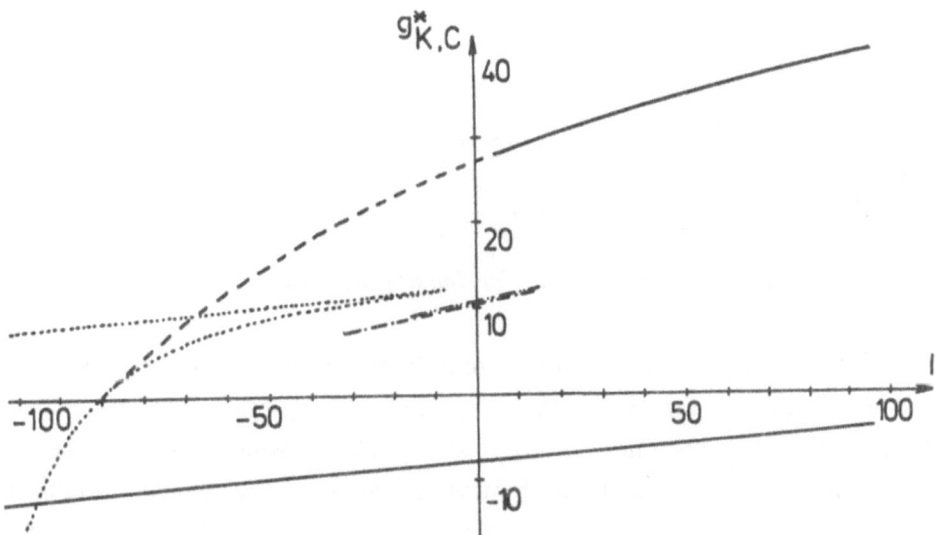

Figure 11. Bifurcation diagram for Chay system obtained using PATH. Solid are super-, dashed are sub-critical Hopf bifurcations, dash-dotted are period doubling bifurcation. The dotted curves enclose a region of multiple equilibria.

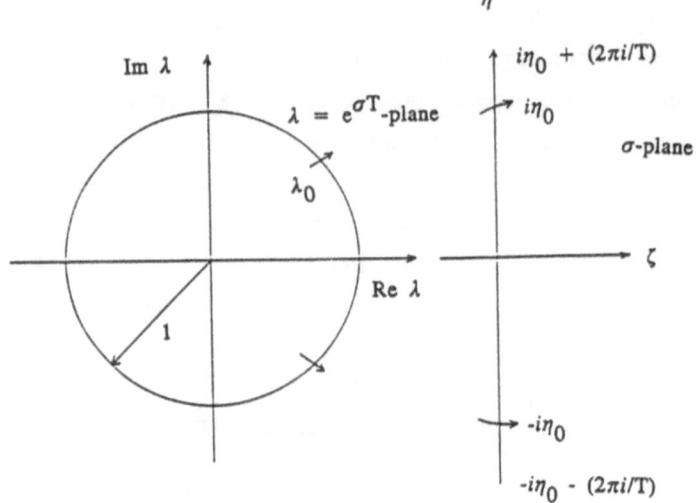

Figure 12. Floquet multipliers and Floquet exponents. Repeated points $(i\eta_0 + 2\pi i k/T)$, $k \in \mathbb{Z}$, on the imaginary axis of the σ plane map into unique points of the complex λ plane.

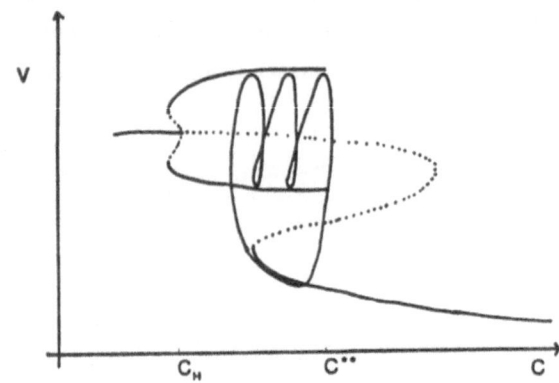

Figure 13. Proposed mechanism for burst generation.

$[Ca^{++}]_i$, and when this reaches some level the burst is terminated. This suggests that the $[Ca^{++}]_i$ acts as a slow variable that modulates the dynamics of the V-n system, which has oscillatory solutions arising at a Hopf bifurcation at C_H and which terminate by coallescing with a saddle point, giving a homoclinic orbit [36]. Thus the addition of an action potential to a burst is not by a bifurcation.

4. Polynomial caricatures.

Before a voltage clamp description of the mechanisms of

the electrophysiology of isolated nerve trunks: these date back to Helmholtz. These observations were interpreted by phenomenological equations, which captured the essential features of the behaviour - a threshold for action potential initiation, and the possibility of repetitive activity.

Phenomenological equations such as the FitzHugh-Nagumo equations [5] are still used in simulations of propagation activity in complicated geometries, such as two and three dimensional excitable media [37].

4.1 The FitzHugh-Nagumo equations

For some purposes it is useful to have a model of an excitable membrane that is mathematically as simple as possible, even if experimental results are reproduced less accurately. Such a model is useful in explaining the general properties of membranes. The use of a qualitative rather than a quantitative model to represent a nonlinear system goes back to Van der Pol. A useful two variable model of an excitable membrane can be developed from the cubic Bonhoeffer-Van der Pol equations.

The cubic FitzHugh-Nagumo equations are:

$$dV/dt = V - V^3/3 - W + I$$

$$(11)$$

$$dW/dt = \Phi(V + a - bW)$$

with a, b and Φ positive parameters. V represents the membrane potential, W a recovery variable and I an injected current. Large amplitude limit cycles arise in the FitzHugh-Nagumo equations *via* Hopf bifurcation into small amplitude, unstable periodic solutions.

4.2 The Hindmarsh-Rose equations.

The cubic term of the FitzHugh-Nagumo equations was introduced to provide a separatrix; another approach is to introduce a cubic to fit momentary current-voltage relations [4] obtained during voltage clamp experiments, as in the Hindmarsh-Rose [39] equations, which were subsequently developed into a three-variable system to account for bursting activity [40]:

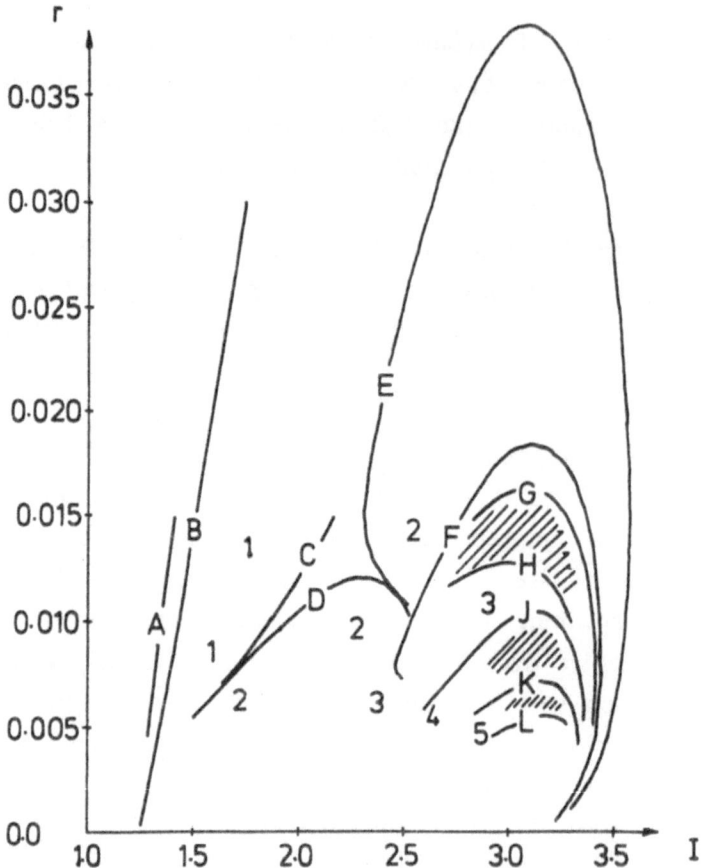

Figure 14. Bifurcation curves for the Hindmarsh-Rose equations, with r=0.015, b=3.0,a=1,c=1,s=4 and x_1=3.

$$dx/dt = y - ax^3 + bx^2 - z + I$$
$$dy/dt = c - dx^2 - y \qquad\qquad (12)$$
$$dz/dt = r(s(x-x_1) -z)$$

where x represents the membrane potential, y a recovery variable and z an adaptation variable, and I a steady applied current. Bifurcation curves obtained using PATH in the I-r parameterplane are presented in Figure 14. The rich behaviour of this simple system is matched by the richness of its bifurcation diagram.

5. Identification of bifurcation points.

Although at a bifurcation point qualitative changes occur in the behaviour of the system, not all changes in qualitative behaviour need be associated with bifurcations. We are interested in identifying bifurcations points:

o in order to compare models with experimental observations

o in order to know how close is a particular periodic discharge to a bifurcation point; this will determine whether or not the periodicity can be annihilated by appropriate perturbations, whether or not transients will be long-lived, exhibiting the critical slowing down seen near phase transitions, and will strongly influence the effects of electrotonic coupling of neurones on synchronisation and entrainment.

When a complete voltage clamp description of the neurone is available, bifurcation curves for the model can be obtained numerically. What we want is to identify bifurcation points from experimental recordings of V(t).

5.1 Supercritical Hopf bifurcation

Small amplitude, almost sinusoidal periodic oscillations will be seen close to a supercritical Hopf bifurcation; the dependence of the amplitude of these oscillations as the root of the distance from where they vanish (at the bifurcation point) is good evidence for a supercritical Hopf bifurcation - see Figure 5.

5.2 Subcritical Hopf bifurcation and phase resetting

The repetitive discharge of the H-H membrane to constant currents represents the activity of a differential system with a limit cycle embedded in an attractor basin. Best [21] examined the response to voltage perturbations during this limit cycle behaviour, measured as phase-response curves, to identify the null space where there is no flow towards or around the limit cycle. A voltage perturbation that takes the system through access portals into the null space will block the repetitive discharge even though the constant current persists: this behaviour is seen in numerical calculations and has been observed in squid axons. Such a null space, with dimension > N-2 , will exist for any dynamical system which exhibits limit-cycle behaviour in an N-dimensional space [41].

If we do not have the mathematical equations that model the neuronal system, then we may study the phase resetting characteristics of the system experimentally. A mathematical representation is then obtained in the form of phase transition curves or phase response curves [42,43].

There are two different types of phase transition curves, namely type 0, type 1, where the average value of the slope is zero or one ' respectively. It is known that, if as the size of the perturbation is changed, the phase transition curve changes from type 0 to type 1, then annihilation of activity by an appropriately timed perturbation of an appropriate amplitude is possible *i.e* there are both a stable equilibrium and a stable (large amplitude) periodic solution possible at this parameter. These two stable solutions will be separated by an unstable periodic solution and this small amplitude, unstable periodic solution has almost invariably emerged *via* a subcritical Hopf bifurcation.

However, whenever periodic solutions emerge from an equilibrium it is almost invariably true that this will be *via* a Hopf bifurcation, as the other routes require a complex conjugate pair of eigenvalues and one or more real eigenvalues to cross the imaginary axis concurrently.

5.3 Lyapunov exponents.

The Lyapunov exponent is a measure of the rate of orbital divergence or convergence in phase space. For a one-dimensional system, if the initial separation of a pair of nearby orbits is $p_i(0)$, after a time of t, the separation becomes $p_i(t)$, then we have the Lyapunov exponent [44,45,46]:

$$\lambda = \lim_{t \to \infty} \frac{1}{t} \log_2 \frac{p_i(t)}{p_i(0)} \tag{13}$$

The Lyapunov exponent is one of the most important quantities to characterize the motion of a dynamical system. Any dynamical system containing at least one positive Lyapunov exponent is defined to be chaotic. The magnitude of a positive Lyapunov exponent reflects the time scale on which system dynamics become unpredictable, and the magnitude of the Lyapunov exponent reflects the time scale on which transients or perturbations of the system state will decay.

A sequence of period-doubling bifurcations provides a well

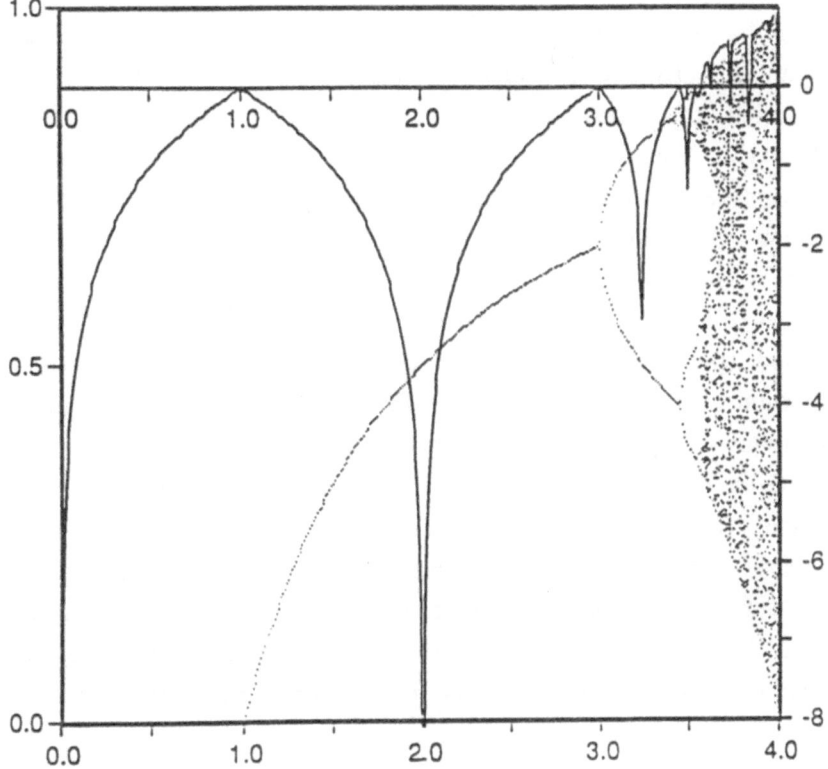

Figure 15. Bifurcation diagram and Lyapunov exponent for quadratic map: at each bifurcation point the Lyapunov exponent kisses zero.

known route into chaos; this can be illustrated by the quadratic interval map:

$$x_{i+1} = ax_i(1 - x_i) = F(x_i)$$

for $0 \le x_i \le 1$ and $0 < a \le 4$; see [47]. This one-dimensional map has a single Lyapunov exponent λ that can be evaluated from

$$\lambda = \lim_{n \to \infty} \sum_{i=1}^{n-1} \log_2 F'[x_i]$$

A bifurcation diagram, with the Lyapunov exponent superposed, is presented in Figure 15. At each bifurcation point, the transcritical bifurcation at $a=1$, the flip bifurcation at $a=3$ and the subsequent infinite sequence of flip (period-doubling) bifurcations at values of a that converge to a limiting point a_∞ in a geometric progression

$$a_k = a_\infty - c \, \mathcal{F}^{-k},$$

where $a_\infty \cong 3.569946$, $c \cong 2.6327$ and $\mathscr{F} \cong 4.6692$ is Feigenbaum's constant, the Lyapunov exponent comes up, touches (but does not cross), and then leaves zero *i.e.* kisses zero. Thus a bifurcation point may be identified by the zero value of the Lyapunov exponent. For a three variable differential system there will be three Lypaunov exponents, (one associated with each eigenvalue) and the Lyapunov spectrum is usually ordered with $\lambda_1 \geq \lambda_2 \geq \lambda_3$. If there is a point attractor the spectrum is (-,-,-); if there is a limit cycle the spectrum is (0,-,-) and so a limit cycle emerges from an equilibrium when the largest Lyapunov exponent becomes 0. Bifurcations between limit cycles (*e.g.* a period doubling bifurcation) are between Lyapunov spectra (0,-,-) and (0,-,-) via a spectrum (0,0,-) at the bifurcation point; the largest negative Lyapunov exponent comes up and kisses zero. Thus if one could estimate Lyapunov exponents as a parameter varies one would be be able to identify bifurcation points. Techniques have been developed for estimating Lyapunov exponents from experimental data.

5.3.1 Estimation of Lyapunov exponent from data

For a differential equation

$$\frac{d}{dt}\vec{x} = F(\vec{x}), \text{ with } \vec{x} = (x_1, x_2, ..., x_n) \tag{13}$$

the solution of this equation takes the following form

$$\vec{x}^0(t) = \phi^t(\vec{x}_0), \tag{14}$$

where $\vec{x}_0$ is the initial condition.

In the neighbourhood of $\vec{x}_0(t)$, any point $\vec{x}(t)$ can be described by its Taylor expansion

$$\vec{x}(t) = \vec{x}_0(t) + \delta\vec{x}(t). \tag{15}$$

Then we have a linear non-autonomous equation:

$$\frac{d}{dt}\delta\vec{x}(t) = U(t)\delta\vec{x}(t) \tag{16}$$

with

$$U(t) = \left[\left. \frac{\partial F_i}{\partial x_j} \right|_{\vec{x}(t)=\vec{x}_0(t)} \right] \qquad (17)$$

U(t) is called the evolution operator or Lyapunov matrix.

If there is a perturbation at $\vec{x}_0$, say $\delta\vec{x}_0$, then after a time t, this perturbation will have an error $\delta\vec{x}(t)$ on $\vec{x}(t)$, and $\delta\vec{x}(t)$ has the form

$$\delta\vec{x}(t) = D\phi^t(\vec{x}_0)\delta\vec{x}_0 \qquad (18)$$

$D\phi^t(\vec{x}_0)$ is a $n\times n$ matrix. Equation (18) reflects the vector change in the tangent space.

Suppose $\vec{e}^i$ (i=1, 2, ..., n) is a set of basis vectors in the tangent space, then if there is the limit

$$\lambda_i = \lim_{t\to\infty}\frac{1}{t} \log_2 \left|\left| D\phi^t(\vec{x}(\vec{x}_0)\vec{e}^i \right|\right| ; \qquad (19)$$

then λ_i is called the Lyapunov exponent, with $||.||$ denoting the vector's norm.

Usually, it is very difficult to use equation (19) to calculate the Lyapunov exponents directly and sometimes it is impossible, and hence we must seek a practical method.

The ODE method was developed independently by the Benettin and Shimada groups [44,49]. For a dissipative dynamical system, we monitor the long-term evolution of a n-dimensional sphere in the n-dimensional phase space to calculate the Lyapunov exponents. First, we determine an arbitrary point as the initial point of the fiducial trajectory produced by integrating the nonlinear equation. Second we construct an orthonormal vector frame anchored to the fiducial trajectory and evolve it with the linearized equation (16). Since each vector diverges in magnitude and tends to fall along the local direction of the most rapid growth, we must use the Gram-Schmidt Re-orthonormalization (GSR) procedure to re-orthonormalize the orthonormal vector frame during its evolution.

Given a set of vectors, after a time of evolution by the equation of (16), we will get a set of new vectors $\{\vec{v}_1, \vec{v}_2, \ldots\ldots, \vec{v}_n\}$. The GSR procedure will provide us with the following orthnormal vectors $\{\vec{v}_1', \vec{v}_2', \ldots\ldots, \vec{v}_n'\}$, which take the form:

$$\vec{v}_1' = \frac{\vec{v}_1}{||\vec{v}_1||}$$

$$\vec{v}_2' = \frac{\vec{v}_2 - <\vec{v}_2, \vec{v}_1'> \vec{v}_1'}{||\vec{v}_2 - <\vec{v}_2, \vec{v}_1'> \vec{v}_1'||} \tag{20}$$

$$\ldots\ldots$$

$$\vec{v}_n' = \frac{\vec{v}_n - <\vec{v}_n, \vec{v}_{n-1}'> \vec{v}_{n-1}' - \ldots\ldots - <\vec{v}_n, \vec{v}_1'> \vec{v}_1'}{||\vec{v}_n - <\vec{v}_n, \vec{v}_{n-1}'> \vec{v}_{n-1}' - \ldots\ldots - <\vec{v}_n, \vec{v}_1'> \vec{v}_1'||}$$

where $<\ >$ signifies the inner product.

From Equation (20), we can see that the GSR procedure never affects the direction of the first vector, so this vector tends to seek out the most rapidly growing direction. The GSR procedure continually changes the second vector's direction, so the second vector is not free in selection of direction. Note however, the vectors $\vec{v}_1'$ and $\vec{v}_2'$ span the same two dimensional subspace as the vectors $\vec{v}_1$ and $\vec{v}_2$, hence this subspace still seeks out the most rapidly growing subspace in spite of repeated replacement of the second vector. The area defined by these two vectors is proportional to $2^{(\lambda_1 + \lambda_2)}$.

The length of vector $\vec{v}_1$ is proportional to $2^{\lambda_1 t}$, so we can calculate λ_1 directly by monitoring the length growth of vector $\vec{v}_1$. Similarly, we can also calculate $\lambda_1 + \lambda_2$ from the area growth of the subspace spanned by $\vec{v}_1$ and $\vec{v}_2$. In practice, since $\vec{v}_1$ and $\vec{v}_2$ are orthogonal, we may determine λ_2 directly from the mean growth rate of the projection of $\vec{v}_2$ on $\vec{v}_2'$. For the same reason, we may determine λ_k directly from the mean growth rate of the projection of $\vec{v}_k$ on $\vec{v}_k'$.

We have illustrated straightforward methods of calculating the Lyapunov exponents for a system whose equation is known. However, the problem is to calculate the Lyapunov exponents (all of them) as parameters are changed, when all we have is a time series recording of neural activity.

Recent research shows great progress in calculating Lyapunov exponents from time series [48 -51]. In this section, we will discuss two different useful method of calculating Lyapunov exponents from an observed time series of single variables.

1. Matrix method

As shown above, we can use a linearized non autonomous equation to describe a vector's change in tangent space

$$\frac{d}{dt}\, \delta\vec{x}(t) \;=\; U(t)\delta\vec{x}(t) \tag{21}$$

Where $U(t)$ takes the form

$$U(t) \;=\; \left[\left. \frac{\partial F_i}{\partial x_j} \right|\; \vec{x}(t)=\vec{x}^0(t) \right] \tag{22}$$

The solution of the equation (21) takes the form

$$\delta\vec{x}(t) \;=\; D\phi^t(\vec{x}_0)\delta\vec{x}_0 \tag{23}$$

Where $D\phi^t$ is an $n \times n$ matrix.

If $\vec{e}_i$ is the base vector of the tangent space, and the limit

$$\lambda_i \;=\; \lim_{t\to\infty} \frac{1}{t}\, \log_2 || D\phi^t(\vec{x}_0)\vec{e}_i || \tag{24}$$

exists, λ_i is called the i-th Lyapunov exponent of the system.

In practice, we divide the time t into k intervals, that is $t=k\Delta t$. The evolution matrix $D\phi^t(\vec{x}_0)$ can be written as

$$
\begin{aligned}
D\phi^t(\vec{x}_0) &= D\phi^{k\Delta t}(\vec{x}_0) \\
&= D\phi^{\Delta t}(\vec{x}_{k-1})D\phi^{\Delta t}(\vec{x}_{k-2})\ldots\ldots D\phi^{\Delta t}(\vec{x}_0).
\end{aligned}
\tag{25}
$$

We use Householder's QR decomposition method: each invertible $n \times n$ matrix can split uniquely into a product of a upper triangular matrix R with non-negative diagonal elements and orthogonal matrix Q, such that

$$D\phi^{\Delta t}(\vec{x}_i) \;=\; Q_{i-1} \cdot R_{i-1} \,. \tag{26}$$

Hence

$$D\phi^{k\Delta t}(\vec{x}_0) = Q_k (R_k R_{k-1} \cdots R_1). \tag{27}$$

We apply equation (27) to equation (24), to give

$$\lambda_i = \lim_{t \to \infty} \frac{1}{kt} \sum_{i=1}^{k} \log_2 r_{jj}^i \tag{28}$$

Where r_{jj}^i is the $(j \times j)$ element of the matrix R_j.

In terms of experimental observations, to calculate the Lyapunov exponent we calculate the matrix $D\phi^t(\vec{x}_i)$, using a time delay method to reconstruct the phase space, and then use a "least-square-error" method to construct it.

Given a series of experimental signals, $\{x_1, x_2, \ldots, x_n\}$, where $x_i = x(i\tau)$, τ is the time interval, we can reconstruct the n-dimensional space

$$\vec{x}_i = (x_i, x_{i+1}, \ldots, x_{i+n-1}), \tag{29}$$

$\vec{x}_i$ is a point in the phase space. We define the distance between two points in the phase space as

$$d^2(\vec{x}_i, \vec{x}_j) = \sum_{k=0}^{n} (x_{i+k} - x_{j+k})^2 \tag{30}$$

if $d \le \varepsilon$, then we say that $\vec{x}_j$ is in the ε-neighbour of $\vec{x}_i$.

Now we examine a selected point $\vec{x}_i$ and other n points in its ε-neighbour, $\vec{y}_j$. The difference vector between $\vec{x}_i$ and $\vec{y}_j$ is defined as

$$\Delta \vec{y}_j = \vec{y}_j - \vec{x}_i \qquad (j=1, 2, \ldots, n) \tag{31}$$

After a time Δt, $\vec{x}_i$ evolves into $\vec{x}_i{}'$, while $\vec{y}_j$ evolves into $\vec{y}_j{}'$, then we get n new difference vectors

$$\Delta \vec{y}_j{}' = \vec{y}_j{}' - \vec{x}_i{}' \qquad (j=1, 2, \ldots, n) \tag{32}$$

If there is an A_i satisfying the following condition

$$\min_{A_i} S = \min \frac{1}{n} \sum_{j=1}^{n} ||\Delta \vec{y}_j{}' - A_i \Delta \vec{y}_j|| \tag{33}$$

then the evolution matrix $D\phi^{\Delta t}(\vec{x}_i)$ can be approximated by A_i.

A_i can be solved from equation (33) by the "least-square-error" method and has

$$A_i = CV^{-1} \tag{34}$$

With C, V being $n \times n$ matrices and

$$c_{kl} = \frac{1}{n} \sum_{j=1}^{n} \Delta y'_{jk} \Delta y_{jl}, \quad v_{kl} = \frac{1}{n} \sum_{j=1}^{n} \Delta y_{jk} \Delta y_{jl} \tag{35}$$

Where Δy_{jk} is the k-th element of $\Delta \vec{y}_k$, $\Delta y'_{jk}$ is the k-th element of $\Delta \vec{y}_j{}'$.

2 The ODE method

Wolf and his colleagues inspired by the ODE method, have developed a method which is conceptually similar to the ODE method. Instead of calculating the eigenvalue of the evolution matrix, we monitor the orbit's divergence directly.

In phase space, we monitor the evolution of a fiducial orbit and one of its nearby orbits, and examine the divergence between them. We select $\vec{x}_0$ as the initial point on the fiducial orbit, and $\vec{x}_1$ as a point in the reference orbit. Initially, the separation between them is $L(t_0)$, after a time of t, the separation becomes $L'(t_1)$. Since in the evolution process, sometimes the fiduciary orbit and the reference orbit fold, sometimes they diverge, so the separation between them sometimes shrinks, sometimes magnifies. We should chose a suitable evolution time that before the two orbits pass through the folding region we change the reference point. Otherwise we may underestimate the value of the Lyapunov exponent. The new reference point should meet two criteria: its separation $L(t_1)$, from the evolved fiducial point should be small, and the angular separation between the evolved previous reference point and the new reference point should be small. We repeat this procedure until the fiducial orbit has traversed the entire data file. Then we get

$$\lambda_1 = \frac{1}{mt} \sum_{k=1}^{m} \frac{L'(t_k)}{L(t_{k-1})} \qquad (36)$$

Where m is the total number of replacement steps.

The algorithm for estimating $\lambda_1 + \lambda_2$ is similar to the proceeding algorithm, but is more complicated in implementation. A triple of points in phase space are chosen, consisting of the initial fiducial point and other two reference points. Evolve these three points until "problems" rise.

The problems include:

o a principle axis vector grow tool large or too rapidly.

o the area grows too rapidly, and the skewness of the area element exceeds a threshold value.

Whenever any of these criteria are met, the triple is evolved backwards and a replacement is attempted. Propagation and replacement steps are repeated until the fiducial orbit has travelled through the entire data file at when we estimate

$$\lambda_1 + \lambda_2 = \frac{1}{mt} \sum_{k=1}^{m} \frac{A'(t_k)}{A(t_{k-1})} \qquad (37)$$

where m is the total number of replacement, $A(t_{k-1})$ is the area spanned by the triple before evolution at step k, and $A'(t_k)$ is the area spanned by the triple after evolution at step k.

5.3.2. Lyapunov exponents for the Hindmarsh-Rose System

To illustrate the principle of identifying bifurcation points we have calculated the Lyapunov exponents for the Hindmarsh-Rose system, where we know there is a rich sequence of bifurcations. Figure 16 shows the Lyapunov exponments for the Hindmarsh-Rose system estimated using the method described following equation (20) from the equations (not from a time series obtained by numerical integration). The second and third Lyapunov exponents have been multiplied by 0.05 and 0.005, and bifurcations can clearly be identified in the range I=1.5 to I= 4 as the second Lyapunov exponent comes up and kisses zero. The three bands of chaos are also evident from the positive first Lyapunov exponent.

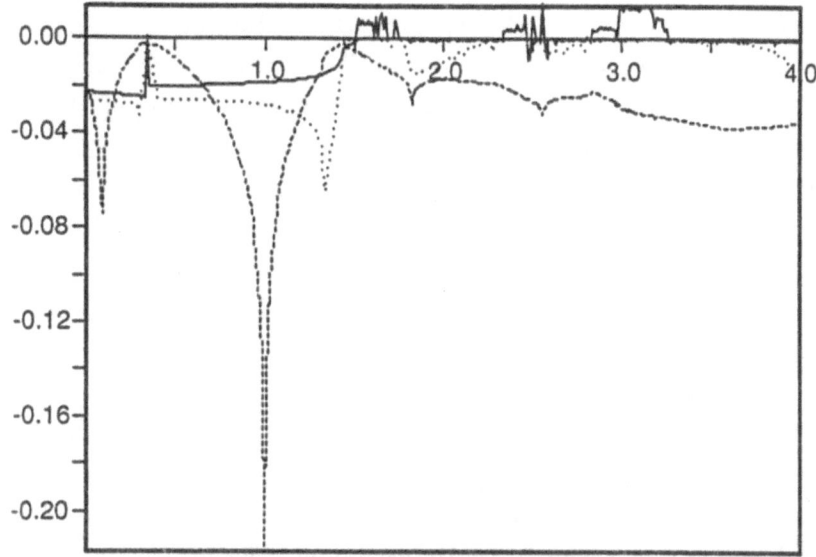

Figure 16. Lyapunov exponents for Hindmarsh Rose system as I is increased from 0 to 4. The second Lyapunov exponent kisses the zero axis at several bifurcation points.

In principle, one should be able to estimate all the Lyapunov exponents from a time series recording of neuronal activity, by the methods described in section 5.3.1. However, although these methods have been evaluated for chaotic pseudo-data, there are problems in applying them to periodic signals (as there is little convergence available onto the attractor), and the presence of noise introduces a positive bias into the estimated Lyapunov exponents. From the estimated Lyapunov exponents of a noisy periodic signal one can confirm a bifurcation point by the second Lyapunov exponent passing through zero; but without other information one could not identify the bifurcation points unequivocally.

6. Bifurcations, synchronisation and binding.

From the viewpoint of the neurobiology of single cells, a neurone may have a repetitive, periodic discharge, and this may change during normal behaviour. Whether or not these changes occur via bifurcations or not does not influence the behaviour generated by the activity. However, the nervous system is a system of interacting cells, and the response of a repetively active cell to inputs from other cells, or to coupling with other cells, will depend on whether or not not the cell is close to a bifurcation point. There is recent interest in the synchronisation of bursting activity of neurones during visual binding [54,55] and its relation to the neuronal basis of attentional mechanisms [56], and we believe that a detailed examination of the behaviour of coupled oscillatory neurones, close to and far from bifurcation points, will provide a firm mathematical basis for interpreting synchronisation of bursting activity in real neural systems.

Ackowledgements. J.Hyde is supported by an S.E.R.C. Mathematics Committee earmarked Mathematical Biology studentship, M.A.Muhamad by an Association of Commonwealth Universities postdoctoral fellowship, and H.G.Zhang by a University of Science and Technology of China studentship. This work is partly supported by a Cognitive Science Initiative Special Program Grant administered by the M.R.C., SPG9017859.

References.

[1] A.V.Holden and W.Winlow. Neuronal activity as the behaviour of a differential system. *I.E.E.E. Trans SMC*-13 711-9 1983

[2] A.V.Holden, J.V.Tucker and B.C.Thompson. The computational structure of neural systems. In: *Neurocomputers and Attention. Vol. 1. Neurobiology, synchronisation and chaos.* A.V.Holden and V.I.Kryukov (eds.). Manchester University Press, Manchester 1991.

[3] A.V.Holden. The mathematics of excitation. In *Biomathematics in 1980.* L.M.Ricciardi & A.Scott (ed.) North-Holland, Amsterdam, 1982

[4] A.V.Holden and W.Winlow. The comparative neurobiology of excitation.In *The neurobiology of Pain.* A.V.Holden and W.Winlow (eds). Manchester University Press, Manchester 1986.

[5] R. FitzHugh. Mathematical models of excitation and propagation in nerve. In *Biological Engineering*, H.P. Schwan (Ed). McGraw-Hill: New York. 1967.

[6] A.V Holden, W. Winlow and P.G. Haydon. Effects of tetraethylammonium and 4-aminopyradine on the somatic potentials of an identified molluscan neuron. *Comp. Biochem. Physiol.* **73A** 303-310 (1982).

[7] A.V. Holden and W. Winlow. Bifurcation of periodic activity from periodic activity in a molluscan neurone. *Biol. Cybern* **42** 189-194 (1982).

[8] W.B. Adams and J.A. Benson. The generation and modulation of endogenous rhythmicity in the *Aplysia* bursting pacemaker neurone R15. *Prog. Biophys. molec. Biol.* **46** 1-49 (1985).

[9] A.L.Hodgkin and A.F.Huxley. Currents carried by sodium and potassium ions through the membrane of the giant axon of *Loligo*. *J.Physiol.***116** 449-472 (1952)

[10] A.L.Hodgkin and A.F.Huxley. The components of membrane conductance in the giant axon of *Loligo*. *J.Physiol.* **116** 473-496 (1952)

[11] A.L.Hodgkin and A.F.Huxley. The dual effect of membrane potential on sodium conductance in the giant axon of *Loligo*. *J.Physiol.***116** 497-506 (1952)

[12] A.L.Hodgkin and A.F.Huxley. A quantitative description of membrane current and its application to conduction and excitation in nerve. *J.Physiol.* **117** 500-544 (1952)

[13] R.G.Casten, H.Cohen and P.A.Lagerstrom. Perturbation analysis of an approximation to the Hodgkin-Huxley. *Quart. Appl. Math.* **32** 365-402 (1975).

[14] G.A. Carpenter. Bursting phenomena in excitable membranes. *SIAM J. Appl. Math.* **36**, 334-372 (1979).

[15] J.Cronin. *Mathematics of cell electrophysiology.* Marcel Decker, New York (1981).

[16] W.C. Troy. The bifurcation of periodic solutions in the Hodgkin-Huxley equations. *Quart. Appl. Math.* **36**, 73-83 (1978).

[17] W.C. Troy. Oscillation phenomena in Hodgkin-Huxley equations. *Proc. Roy. Soc. of Edinburgh.* **74A** 299-310 (1976).

[18] W.C. Troy. Large amplitude periodic of a system of equations derived from Hodgkin-Huxley equations. *Arch. Rational Mech. Analysis* **65** 227-247 (1976).

[19] B.D. Hassard. Bifurcation of periodic solutions of the Hodgkin-Huxley model for the squid giant axon. *J. Theor. Biol.* **71** 401-420 (1978).

[20] B.D. Hassard, N.D. Kazarinoff and Y.H. Wan. *Theory and application of Hopf bifurcation.* Cambridge, England: Cambridge Univ. Press, 1981.

[21] E.N. Best. Null space in the Hodgkin-Huxley equations: A critical test. *Biophs. J.* **27** 87-104 (1979).

[22] R. Guttman, S. Lewis and J. Rinzel. Control of repetitive firing in squid axon membrane as a model of a neurone oscillator. *J. Physiol.* **305**, 377-395 (1984).

[23] J. Brindley, C.Kaas-Petersen and A.Spence Path-following methods in bifurcation problems. *Physica-D* **34** 456-461 (1989).

[24] E.J. Doedel. AUTO-Software for continuation and bifurcation problems in ordinary differential equations. *Calif. Inst. Tech.* (1986).

[25] C.K. Peterson. *Techniques to trace curves of bifurcation points of periodic solutions.* Path Users Guide. Centre for Nonlinear Studies, Leeds. 1987.

[26] N. Chalazonitis and M.Boisson (eds) *Abnormal neuronal discharges.* (1978).

[27] A.V.Holden, M.A.Muhamad and A.K.Schierwagen. Repolarising cuurents and periodic activity in nerve membrane. *J.theor. Neurobiol.* **4** 61-71 (1985).

[28] I.S.Labouriau. In *Choas in biological systems.* H.Degn, A.V.Holden and L.F.Olsen (eds). Plenum, New York 105-112 (1987).

[29] A.V. Holden and M. Yoda. Bifurcation theory and autorhythmicity of the excitable membrane on nerve cells. *Proc. 2nd World Congr. on mathematics on the service of man.* A. Ballester, D. Cardus and E. Trillas (eds). Las Palmas: Universidad Politecnica. 355-360 (1982).

[30] A.V. Holden, P.G. Haydon and W. Winlow.Multiple equilibria and erotic behaviour in excitable membranes. *Biol. Cyber.* **46** 167-182 (1983).

[31] P.G.Kostyuk, N.S.Veselovsky, S.A.Fedulova and A.Y. Tsyndrenko. Ionic currents in the somatic membrane of rat dorsal root ganglion cells. *Neuroscience* **6** 2423-2444 (1981).

[32] T.R.Chay. Chaos in a three variable model of an excitable cell.*Physica-D* **16** 233-242 (1985).

[33] T.R.Chay and J.Rinzel. Bursting, beating, and chaos in an excitable membrane model. *Biophys. J.* **47** 357-366 (1985).

[34] T.R.Chay and H.S.Kang. Multiple oscillatory states and chaos in the endogenous activity of excitable cells. In: *Chaos in Biological Systems.* H.Degn, A.V.Holden and L.F.Olsen (eds) Plenum,

New York (1986).

[35] J.M.T. Thompson and H.B. Stewart. *Nonlinear dynamics and chaos.* John Wiley and Sons Ltd. 1986.

[36] G. Iooss and D.D. Joseph. *Elementary stability and bifurcation theory.* New York: Springer-Verlag. 1980.

[37] J. Rinzel and W.C. Troy. A one-variable map analysis of bursting in the Belousov-Zhabotinskii reaction. In *Nonlinear partial differential equations.*J.A. Smoller (ed). American mathematical society, Providence. 411-428 (1983).

[38] A.V. Holden, M. Markus and H.G. Othmer (eds) *Nonlinear wave processes in excitable media.* Plenum: New York. 1991.

[39] J.L.Hindmarsh and R.M.Rose. A model of the nerve impulse using two first-order differential equations. *Nature* **296** 162-3 (1982).

[40] J.L.Hindmarsh and R.M.Rose. A model of neuronal bursting using three coupled first order differential equations. *Proc. Roy. Soc.Lond.* **B221** 87 (1984).

[41] J. Guckenheimer. Isochrons and phaseless sets. *J. math. Biol.***1** 259-273 (1975).

[42] M. Barbi, P. Haydon, A.V. Holden and W. Winlow. On the phase response curves of repetitively active neurones. *J. Theor. Neurobiol.* **3**, 15-24, (1984).

[43] T. Pavlidis. *Biological oscillators: their mathematical analysis.* Academic Press. 1973.

[44] G.Bennetin, L.Galgani, A. Giorgilli, J.-M.Strelcyn. Lyapunov characteristic exponents for smooth dynamic system and for Hamiltonian system; method for caculation all of them. *Meccanica,* **15**, 9 (1980).

[45] A.M.Lyapunov. General problem of stability of motion. *GITTL,* Moscow (1950); French transl. *Ann. Math. Study,* Princeton University press. **17**. (1947).

[46] V.I. Oseledec. A multiplicative Ergodic theorem, Lyapunov characteristic number for dynamical system. *Trans. Moscow Math. Soc.,* **19**, 197 (1968).

[47] H.Lauwerier. One dimensional iterative maps in *Chaos* ed. A.V.Holden. Manchester University Press (1986).

[48] J Holzfuss, W. Lauterbon. Lyapunov exponents from a time series of acoustic chaos. *Phys. Rev.* A, **39** (1989).

[49] I.Shimada, T.Nagashima. A numerical approach to Ergodic

problem of dissipative dynamical systems. *Prog. Theor. Phys.* **61** 1605 (1979).

[50] A.Wolf, J.B.Swift, H.L.Swinney, J.A.Vastano. Determining Lyapunov exponents from time series, *Physica D* **16** (1985)

[51] J.-P.Eckmann, S.O Kamphorst,D.Ruelle, S.Ciliberto, Lyapunov exponents from time series, Phys. Rev. A, 34, 4971 (1986).

[52] H.Nagashima, *J.Phys. Soc., Japan,* **51**, 21 (1982).

[53] N.H.Parkard, J.P.Crutchfield. *Phy. Rev. Lett.* **45**, 712 (1980).

[54] C.M.Gray, P.Konig, A.K.Engel and W.Singer. Oscillatory responses in cat visual cortex exhibit inter-columnar synchronization which reflects global stimulus properties. *Nature* **338** 334-337 (1989)

[55] R.Eckhorn, T.Schanze and H.J.Reitboeck. Neural mechanisms of flexible feature linking in the visual system. In *Mathematical Approaches to Brain Functioning Diagnostics,* I.Dvorak and A.V.Holden (eds). Manchester University Press (1991)

[5] A.V.Holden and V.I.Kryukov (eds) *Neurocomputers and attention. Vol.I: Neurobiology, synchronisation and chaos.* Manchester University Press (1991)

A MODEL FOR LOW THRESHOLD OSCILLATIONS IN NEURONS

J.L.Hindmarsh[1] and R.M.Rose[2]

[1] School of Mathematics and [2]Department of Physiology
Senghennydd Road
P.O.Box No. 915
Cardiff CF2 4AG
United Kingdom

1 Introduction

This work introduces a model which attempts to explain the intracellular recordings made by Wilcox, Gutnick & Christoph (1988) of low threshold oscillations in neurons in the lateral habenula nucleus. Our purpose here is not to discuss all the physiological details but rather to concentrate on the structure of the dynamical system that we think is responsible for the main aspects of the behaviour of these low threshold oscillations. We are looking for a simple, low dimensional system of differential equations that describe a dynamical system whose behaviour is qualitatively similar to that of the neurons studied by Wilcox, Gutnick, & Christoph (1988).

First we will describe briefly those aspects of their observations that we have found interesting. We will then attempt to explain them by using two-dimensional models; this will allow us to introduce, in a simple context, the important ideas before discussing the main model. Finally we will indicate briefly a possible physiological justification for this model as well as speculating on the possible significance of the model.

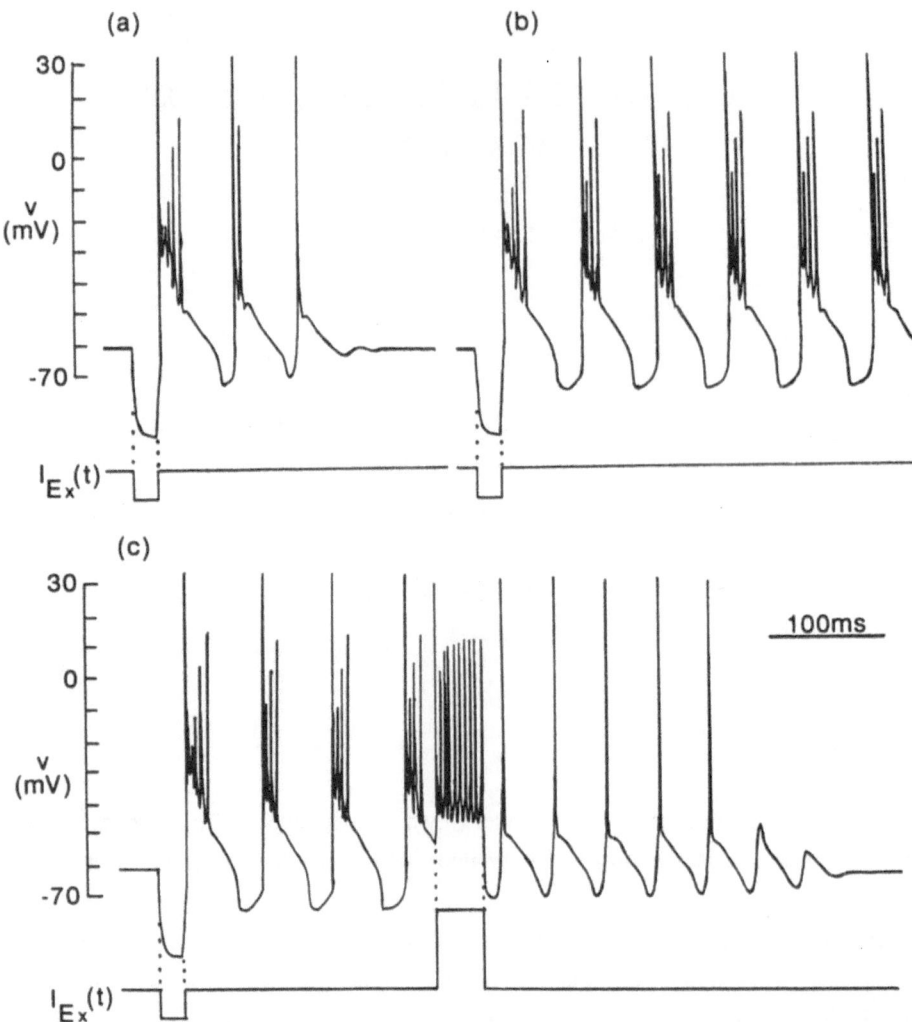

Figure 1. Responses of an eight-dimensionl model of a lateral habenula cell to current steps $I_{Ex}(t)$. The graphs show the membrane potential and external current against time. In response to a negative current step applied to the cell in equilibrium, we obtain either (a) a decaying oscillation, or (b) a prolonged oscillation. Shown in (c) is the same oscillation as in (b) but this time it is terminated by a positive current step.

2 The observations

The particular features for which we wish to build a model are shown in figure 1. This figure is not in fact taken from the paper of Wilcox, Gutnick & Christoph (1988) but comes from an eight-dimensional model that we have developed from our earlier model of a thalamic neuron (Rose & Hindmarsh 1989 a,b,c). We describe a reduced version of the eight-dimensional model in section 4 below, and will present further details in another paper which we are preparing.

The graphs in figure 1 show the membrane potential against time in response to current steps. They were computed using our model, but we will refer to them as though they were actual recordings. We see a low frequency oscillation of the membrane potential between $\approx$ -75mV and $\approx$ -45mV. For membrane potentials above $\approx$ -50mV there is an additional very rapid oscillation, the so called "action potentials", which are the "spikes on the top of the slow wave". We will disregard these action potentials and concern ourselves only with the underlying low threshold oscillation. The features that we wish to draw attention to are the following: when the cell is in equilibrium a negative current step triggers either a low threshold oscillation which decays (figure 1a) or, in some rare cases, a prolonged oscillation (figure 1b) and that this oscillation can be terminated by a positive current step (figure 1c).

3 Two-dimensional models

A simple dynamical system that describes a decaying oscillation such as that shown in figure 1a is given by

$$\left. \begin{array}{l} \dot{x} = -y-\alpha x \\ \dot{y} = x-\alpha y \end{array} \right\} \tag{1}$$

where $\alpha > 0$, whose phase diagram is an anticlockwise inward spiral around the stable equilibrium point at the origin. This is evident using polar coordinates r, θ for points in the x,y plane. With these coordinates equations (1) become

84

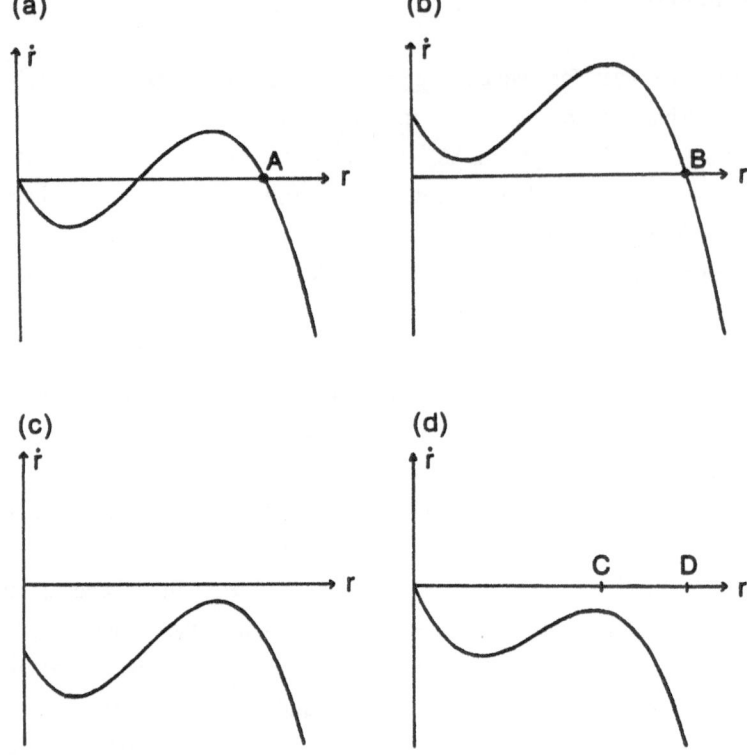

Figure 2. Graphs of $\dot{r}$ against r for $\dot{r} = -\alpha r(r^2 - 2ar + b) - I_{Ex}$ for the cases, (a) $I_{Ex} = 0$, $b < a^2$, (b) $I_{Ex} < 0$, $b < a^2$, (c) $I_{Ex} > 0$, $b < a^2$, and (d) $I_{Ex} = 0$, $b > a^2$.

$$\left. \begin{aligned} \dot{r} &= -\alpha r \\ \dot{\theta} &= 1 \end{aligned} \right\}.$$ (2)

Our interpretation of figure 1b, where the oscillation shows no sign of decaying, is that there is a stable limit cycle present. In order to include this feature we change equations (2) to

$$\left. \begin{aligned} \dot{r} &= -\alpha r(r^2 - 2ar + b) \\ \dot{\theta} &= 1 \end{aligned} \right\}$$ (3)

where α, a, b are all greater than zero. This system has a stable E.P. at $r = 0$ and, provided $b < a^2$, limit cycles at $r = a \pm \sqrt{(a^2 - b)}$, the outer of which is stable. If b is just greater than a^2 the system will just fail to have a limit cycle. This would appear to be the case for most of the cells recorded by Wilcox and colleagues.

The next problem is to explain the effect of external current steps on the system. An external current is taken into account by adding a term $I_{Ex}(t)$ to the expression for $\dot{v}$, see for instance equations (13). Although neither r nor θ is a suitable candidate for v, we will for the moment identify $-r$ with v. Equations (3) then become

$$\left. \begin{aligned} \dot{r} &= -\alpha r(r^2 - 2ar + b) - I_{Ex}(t) \\ \dot{\theta} &= 1 \end{aligned} \right\}$$ (4)

where we will assume that $b < a^2$ to allow a stable limit cycle in the case that $I_{Ex}(t) = 0$.

The equations (4) serve to illustrate a simple mechanism whereby the system can be switched from the stable E.P. to the stable limit cycle, using a negative step and *vice versa* for a positive step. This is easily seen by considering the graphs of $\dot{r}$ against r shown in figure 2.

Suppose initially that $r = 0$, then sufficiently negative $I_{Ex}(t)$ means that $\dot{r} > 0$ and r increases until the point B is reached (figure 2b). If $I_{Ex}(t)$ now returns to 0, then r decreases, not to 0, but to the point A corresponding to the stable limit cycle (figure 2a). The negative step has switched the system from its steady resting state to a steady oscillation. On the other hand starting in this stable limit cycle, the effect of a positive current is to make $\dot{r} < 0$ and r decrease towards 0 (figure 2c). Thus a sufficiently large and long-lasting positive step terminates the oscillations associated with the stable limit cycle and returns the system to its stable E.P..

The identification of r with the membrane potential is unsatisfactory because $r \geq 0$, whereas v is unbounded. We will therefore return to the Cartesian form. The equations are now

$$\left.\begin{aligned} \dot{x} &= -y - \alpha\phi(r)x \\ \dot{y} &= x - \alpha\phi(r)y \end{aligned}\right\} \tag{5}$$

where $\phi(r) = (r^2 - 2ar + b)$, $r = \sqrt{(x^2 + y^2)}$.

If we identify x with the membrane potential v and include the external current $I_{Ex}(t)$, we obtain

$$\left.\begin{aligned} \dot{x} &= -y - \alpha\phi(r)x + I_{Ex}(t) \\ \dot{y} &= x - \alpha\phi(r)y \end{aligned}\right\} \tag{6}$$

As above, we consider the case where the values of a and b permit a stable limit cycle. Once again a sufficiently strong and prolonged current step can switch the system of equations (6) from the stable E.P. at (0,0) to the stable limit cycle. But with this identification of x with v, it does not matter whether the step is positive or negative. Furthermore although the oscillation can be terminated by a suitable step, the timing, as well as the magnitude and duration, is critical. It appears that there was no such difficulty experimentally and so this model for the termination of the oscillation is unsatisfactory. The model of equations (4) and the models below do not suffer from this defect.

Before leaving this simple model we would like to note two additional features. If the observations of Wilcox and colleagues are to be interpreted using a model similar to the one described by equations (4) above then it would seem that most of the neurons they studied would correspond to the case where there was not a stable limit cycle present. This is because of the infrequency of the case where there was a prolonged oscillation. This suggests that the values of a and b should be chosen so that the system just fails to have a stable limit cycle.

With this choice the graph of $\dot{r}$ against r for the case $I_{Ex}(t) = 0$ will be like that shown in figure 2d. Suppose the system is driven from its stable E.P. by a negative current step which increases r to the point D. When the pulse is switched off the system will not be held in a stable limit cycle but will return to the E.P.. The important point is that when r is decreasing near the point C where the limit cycles just failed to appear, $\dot{r}$ may have very small values. This would mean that the oscillations could take a comparatively long time to decay, endowing the cell with an impermanent "memory" of the step. A limit cycle would allow the cell a permanent "memory" of the step.

The second point is that a current step is not the only way to drive the system from its stable E.P. into oscillation, whether held in a stable limit cycle or decaying. We could also, in the case of the model described by equations (6), use periodic current pulses. Whether or not a train of pulses would significantly affect the system would depend on the strength and duration of the individual pulses, the number of pulses and of course their frequency. The most effective frequency would presumably be close to the frequency of the oscillation generated. The cell can be thought of as being tuned to respond to a particular frequency. These two features, which for convenience we will call "tuning" and "memory", will reappear in the more elaborate model of the next section.

4 A three-dimensional model

The model we describe in this section is a simplification of a much more detailed model. As such it should be thought of as a model of a model. Nonetheless we feel that it shows clearly some of the underlying dynamics of the detailed model and, perhaps, of the neurons themselves.

In section 5 we will give a brief description of this simplification. It will involve consideration of a system near an E.P. using real canonical form coordinates. A result of the transformation to these coordinates is that the membrane potential cannot be identified with any particular coordinate. It is instead a linear combination of them. This in turn means that the introduction of the external current looks somewhat unusual.

The equations of the model are

$$\dot{x} = -\varepsilon\alpha[1+a(x^2+y^2)-z]x-y+g(t)\cos\phi$$
$$\dot{y} = -\varepsilon\alpha[1+a(x^2+y^2)-z]y+x+g(t)\sin\phi$$
$$\dot{z} = -\varepsilon\gamma[z-(1+a)(x^2+y^2)+b(x^2+y^2)^2]+g(t)p \tag{7}$$

where ϕ and p are constants and $g(t)$ represents the external current. We will restrict our attention to external currents of the form

$$g(t) = \varepsilon(g_1+g_2\cos\omega t)$$

where g_1,g_2 and ω are constants, with $g_2 \geq 0$ and $\omega = 1+\varepsilon\delta$. In this way we can examine the effects of both current steps and periodic pulses. The parameter δ has been introduced to allow for some measure of detuning.

Using polar coordinates r,θ instead of x,y equations (7) become

$$\dot{r} = -\varepsilon[\alpha r(1+ar^2-z)-(g_1+g_2\cos\omega t)\cos(\phi-\theta)]$$
$$\dot{\theta} = 1+\varepsilon[(g_1+g_2\cos\omega t)\sin(\phi-\theta)/r]$$
$$\dot{z} = -\varepsilon[\gamma(z-(1+a)r^2+br^4)-(g_1+g_2\cos\omega t)p] \tag{8}$$

The parameter ε signals terms that we will regard as small for the purpose of the averaging which follows. Before averaging we put

$$\psi(t) = \omega t-\theta(t)+\phi$$

and the equations become

$$\dot{r} = -\varepsilon\{\alpha r(1+ar^2-z)-g_1\cos(\psi-\omega t)$$
$$-g_2[\cos\psi-\cos(\psi-2\omega t)]/2\}$$
$$\dot{\psi} = \varepsilon\{\delta-g_1(\sin(\psi-\omega t))/r$$
$$-g_2[\sin\psi-\sin(\psi-2\omega t)]/(2r)\} \tag{9}$$
$$\dot{z} = -\varepsilon\{\gamma(z-(1+a)r^2+br^4)-g_1p-g_2\cos\omega t\}$$

After averaging over a time period of $2\pi/\omega$ these equations become

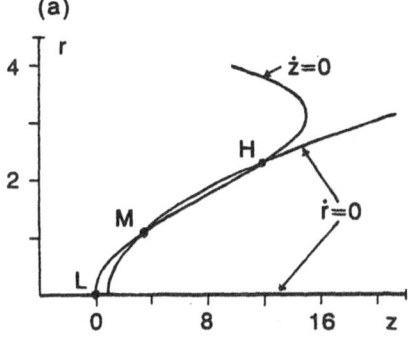

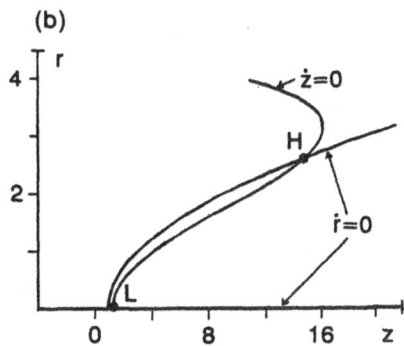

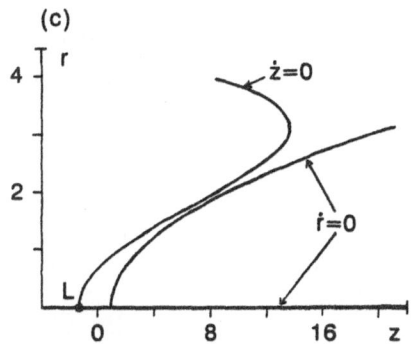

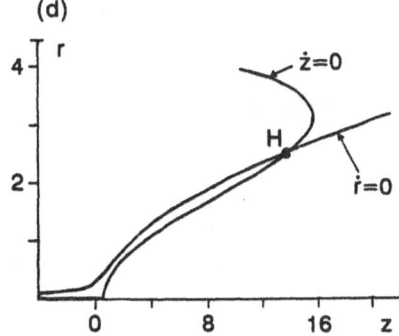

Figure 3. The r and z nullclines for equations (11) are shown for the cases, (a) $g_1 = 0$, (b) $g_1 < 0$, and (c) $g_1 > 0$. The r and z nullclines for equations (12) are shown in (d).

$$\left.\begin{array}{l} \dot{r} = -\varepsilon[\alpha r(1+ar^2-z)-g_2(\cos\psi)/2] \\[4pt] \dot{\psi} = \varepsilon[\delta-g_2(\sin\psi)/(2r)] \\[4pt] \dot{z} = -\varepsilon[\gamma(z-(1+a)r^2+br^4)-g_1p] \end{array}\right\} \qquad (10)$$

We first consider the case where the external current is constant. This means that $g_2=0$, the equation for ψ is decoupled and we need consider only the averaged equations for r and z. These are

$$\left.\begin{array}{l} \dot{r} = -\varepsilon\alpha r(1+ar^2-z) \\[4pt] \dot{z} = -\varepsilon[\gamma(z-(1+a)r^2+br^4)-g_1p] \end{array}\right\} \qquad (11)$$

We can learn much about these equations by looking at the r and z nullclines. The r nullcline consists of points on the z-axis and points satisfying the equation

$$z = 1+ar^2.$$

The z nullcline consists of the points satisfying the equation

$$z = (1+a)r^2-br^4+g_1p/\gamma.$$

Clearly the location of the z nullcline depends on the value of g_1 and the sign of p. In the next section we will see that p is negative. Thus the effect of increasing g_1 is to translate the z nullcline to the left as shown in figure 3. This figure shows the nullclines in the z,r plane for the cases (a) $g_1 = 0$, (b) $g_1 < 0$ and (c) $g_1 > 0$. The equilibrium points, lower, middle and higher, are labelled L,M,H. In figure 3a L and H are stable and M is unstable. In figure 3b L is unstable and H is stable. In figure 3c L is stable. Suppose the system, with $g_1=0$, is in equilibrium at point L of figure 3a. If a negative current step is applied, $g_1 < 0$, then the nullcline diagram changes from figure 3a to figure 3b and the state of the system changes to the E.P. at H in figure 3b. This corresponds to a stable limit cycle for the three-dimensional system. If the current step is now terminated, so $g_1 = 0$, the system changes its state to the stable E.P. at H in figure 3a. Thus the effect of a sufficiently long and prolonged negative current step is to switch the three-dimensional system from its stable E.P. to its stable limit cycle.

Conversely, suppose the system, with $g_1 = 0$, is in the E.P. at H of figure 3a. If a positive current step is applied then the nullcline diagram changes from figure 3a to figure 3c and the state of the system changes to the stable E.P. at L in figure 3c. If the current step is now terminated the system changes its state to the stable E.P. at L in figure 3a. Thus the effect of a sufficiently long and prolonged positive current step is to switch the three-dimensional system from its stable limit cycle to its stable E.P..

We will now consider the case where we have a periodic external current. We take $g_1 < 0$ and $g_2 = |g_1|$ so that the external current is $-\varepsilon|g_1|(1-\cos\omega t)$. This may be regarded as a sinusoidal approximation to a train of negative current pulses of amplitude $2\varepsilon|g_1|$. The effect of this forcing term is frequency dependent but it will be sufficient to discuss the case $\omega = 1$ where $\delta = 0$. In this case $\psi \to 0$ and we are left with the equations

$$\left.\begin{array}{l} \dot{r} = -\varepsilon[\alpha r(1+ar^2-z) - |g_1|/2] \\ \dot{z} = -\varepsilon[\gamma(z - (1+a)r^2+br^4) - g_1 p] \end{array}\right\} \tag{12}$$

whose z and r nullclines are

$$z = 1+ar^2 - |g_1|/(2\alpha r)$$

and

$$z = (1+a)r^2 - br^4 + g_1 p/\gamma$$

respectively and are shown in figure 3d. Both nullclines change position as g_1 varies. As the amplitude, $2\varepsilon|g_1|$, of the periodic external current increases, the z nullcline is translated to the right, as in the constant negative current case, and the r nullcline is raised above its previous position by an amount that depends on z. Both of these changes contribute to moving the lower and middle E.P.'s closer together. For sufficiently large $|g_1|$ these E.P.'s will meet and disappear.

If the system is initially in its stable E.P. with no external current, namely the E.P. L of figure 3b, then a sufficiently large negative periodic current pulse will switch the system into its stable limit cycle. Because both changes of position of the nullclines contribute it is possible to make this switch with a lower current amplitude than the constant current amplitude required. In other words the system will be more sensitive to incoming periodic pulses provided that they are of the right frequency.

We will not discuss the effect of detuning, $\delta \neq 0$, other than to remark that it decreases the sensitivity of the system as would be expected.

We have considered the case which allows a limit cycle when there is no external current. This case is apparently rare so, as remarked in section 3, we would expect most cells to be described by models whose parameters do not allow this possibility. However, because some cells do, we suggest that the parameters are such that the limit cycle only just fails to appear.

In saying that the limit cycle only just fails to appear we mean that the r and z nullclines, that are shown intersecting at M and H in figure 3a, no longer intersect but still lie close together. A consequence of this is that following a negative current step the system is not held in a limit cycle but returns to the stable E.P.. In so doing the phase point representing the state of the system returns through the narrow channel between the nullclines. Being close to both nullclines its movement is slow, which means that the oscillation of the three- dimensional system is slow to decay. This is analogous to the impermanent memory described in section 3.

5 A physiological model

In this section we will describe very briefly what lead us to propose the model of section 4. Although we think that model is valuable in its own right, it is important to provide physiological justification for it.

In Rose and Hindmarsh (1989 a) we started from the Hodgkin-Huxley equations (Hodgkin & Huxley 1952) with the addition of a fast transient outward current I_A (Connor & Stevens 1971). To these equations we added a low threshold transient calcium current I_T (which we referred to as I_{Sa}), so that the equation for the rate of change of the membrane potential was

$$\dot{v} = C^{-1}(-I_{Na}-I_K-I_L-I_A-I_T).$$

In addition to the membrane potential there were six other variables. By a series of approximations we were able to reduce the number of variables leaving us with the three equations

$$
\left.
\begin{aligned}
\dot{v} &= C^{-1}\{-g_{Na}\,m_\infty^3(v)(-3(q-Ab_\infty(v))+0.85)(v-v_{Na})-I_L \\
&\quad -g_K q(v-v_K)-g_T s_{T\infty}(v)h_T(v-v_{Ca})+I_{Ex}(t)\} \\
\dot{q} &= \tau_q^{-1}(v)(q_\infty(v)-q) \\
\dot{h}_T &= \tau_{hT}^{-1}(h_{T\infty}(v)-h_T)
\end{aligned}
\right\}
\tag{13}
$$

A further reduction can be made if we concentrate on the subthreshold region where $q \approx q_\infty(v)$ and ignore the action potential by putting $g_{Na} = 0$. Doing this we are left with the equations

$$
\left.
\begin{aligned}
\dot{v} &= C^{-1}\{-I_L - g_K q_\infty(v)(v-v_K)-g_T s_{T\infty}(v)h_T(v-v_{Ca})+I_{Ex}(t)\} \\
\dot{h}_T &= \tau_{hT}^{-1}(h_{T\infty}(v)-h_T)
\end{aligned}
\right\}
\tag{14}
$$

In these equations h_T is an inactivation variable for the low threshold current I_T. Further details of this procedure may be seen in Rose and Hindmarsh (1989 a,b,c) where it should be noted that we used the terms $I_{Sa}, g_{Sa}, s_{a\infty}(v), h_a$ and τ_{ha} instead of $I_T, g_T, s_{T\infty}, h_T$ and τ_{hT} used here.

In order to explain the low threshold oscillations of lateral habenula cells, we added a calcium activated potassium current

$$
I_{KCa(T)} = g_{KCa(T)}\left(\frac{c}{K_{Ca(T)} + c}\right)(v-v_K)
$$

where $g_{KCa(T)}$ and $K_{Ca(T)}$ are conductivity and dissociation constants respectively and c measures the concentration of intracellular free calcium (see Plant (1978)). The differential equation for c is

$$
\dot{c} = -kI_T-k_{Ca}c
$$

where k and k_{Ca} are constants.

We now have a three-dimensional system

$$\dot{v} = C^{-1}\{-I_L - g_K q_\infty(v)(v-v_K) - g_T s_{T\infty}(v)h_T(v-v_{Ca})$$
$$- I_{KCa(T)} + I_{Ex}(t)\}$$
$$\dot{h}_T = \tau_{hT}^{-1}(h_{T\infty}(v) - h_T) \qquad\qquad (15)$$
$$\dot{c} = -kI_T - k_{Ca} c$$

With suitable choices for the various constants these equations have a stable E.P. in the subthreshold region. This will be explained in more detail in another paper which we are preparing.

To simplify the discussion we define the vector v by

$$v = \begin{bmatrix} v \\ h_T \\ c \end{bmatrix}$$

so that equations (15) may be written as

$$\dot{v} = F(v) + f(t)$$

where F(v) represents the right hand side of equations (15) except for the external current term which is contained in f(t) as

$$f(t) = C^{-1}\begin{bmatrix} I_{Ex}(t) \\ 0 \\ 0 \end{bmatrix}.$$

Let the E.P. be at v_0 so $F(v_0) = 0$. Put $w = v-v_0$ and the equations become

$$\dot{w} = \dot{v} = F(v) + f(t)$$
$$= F(v_0 + w) + f(t) \qquad\qquad (16)$$
$$= Aw + H(w) + f(t)$$

where $A = DF(v_0)$ is the linear approximation matrix and H(w) are the higher order terms. The matrix A has one real eigenvalue $-\gamma$ and a complex conjugate pair $-\alpha \pm i\beta$, where α, β, γ are all positive. Now transform to the real canonical form of A using the transformtion T.

$$T\dot{w} = TAT^{-1}Tw + TH(T^{-1}Tw) + Tf(t)$$

or

$$\dot{x} = A_c x + H_c(x) + Tf(t)$$

where $x = Tw$, A_c is a real canonical form of A and $H_c(x) = TH(T^{-1}x)$. This can be written as

$$\left.\begin{aligned}
\dot{x} &= -\alpha x - \beta y + H_{c1}(x,y,z) + p_1(t) \\
\dot{y} &= \beta x - \alpha y + H_{c2}(x,y,z) + p_2(t) \\
\dot{z} &= -\gamma z + H_{c3}(x,y,z) + p_3(t)
\end{aligned}\right\} \tag{17}$$

where $p(t) = Tf(t)$. For convenience we will introduce ϕ and p by

$$\begin{bmatrix} T_{11} \\ T_{21} \\ T_{31} \end{bmatrix} = \begin{bmatrix} \cos\phi \\ \sin\phi \\ p \end{bmatrix} (T_{11}^2 + T_{21}^2)^{1/2}$$

and put

$$g(t) = (T_{11}^2 + T_{21}^2)^{1/2} C^{-1} I_{Ex}(t)$$

so that

$$Tf(t) = g(t) \begin{bmatrix} \cos\phi \\ \sin\phi \\ p \end{bmatrix}$$

where it is important to note, that for the E.P. under consideration, and with the parameters we are using, p is negative. In section 4 we saw that the sign of p determined the effect a constant external current had on the system. Namely that a negative current switched the oscillation on and a positive step switched it off. Using polar coordinates r,θ instead of x,y, equations (17) become

$$\left.\begin{aligned}
\dot{r} &= -\alpha r + H_{c1}\cos\theta + H_{c2}\sin\theta + g(t)\cos(\phi-\theta) \\
\dot{\theta} &= \beta + (-H_{c1}\sin\theta + H_{c2}\cos\theta + g(t)\sin(\phi-\theta))/r \\
\dot{z} &= -\gamma z + H_{c3} + g(t)p
\end{aligned}\right\} \tag{18}$$

It was the numerical study of equations (18) that lead us to propose equations (7) as a model of the model.

96

6 Conclusion

We believe that the three dimensional model described in section 4 contains the essential dynamical system of the cells in the lateral habenula nucleus. If this is correct then the model suggests that these cells might exhibit the frequency sensitivity to incoming signals described in section 4. This would have interesting implications for the understanding of information processing in the central nervous system.

This work was supported by the Wellcome trust.

One of the authors, (J.L.H.), is grateful to Mr. Michael Quinn for assistance in preparing this paper.

References

Connor,J.A. & Stevens,C.F. 1971 Prediction of repetitive firing behaviour from voltage-clamp data on an isolated neurone soma. *J.Physiol.,Lond.* **213**, 31-53.

Hodgkin,A.L. & Huxley,A.F. 1952 A quantitative description of membrane current and its application to conduction and excitation in nerve. *J.Physiol.,Lond.* **117**,500-544.

Plant,R.E. 1978 The effects of calcium^{++} on bursting neurons. *Biophys.J.* **21**, 217-237.

Rose,R.M. & Hindmarsh,J.L. 1989a The assembly of ionic currents in a thalamic neuron. I The three-dimensional model. *Proc.R.Soc.Lond.*B **237**, 267-288.

Rose,R.M. & Hindmarsh,J.L. 1989b The assembly of ionic currents in a thalamic neuron. II The stability and state diagrams. *Proc.R.Soc.Lond.*B **237**, 289-312.

Rose,R.M. & Hindmarsh,J.L. 1989c The assembly of ionic currents in a thalamic neuron. III The seven-dimensional model. *Proc.R.Soc.Lond.*B **237**, 313-334.

Wilcox,K.S., Gutnick,M.J. & Christoph,G.R. 1988 Electro-physiological properties of neurons in the lateral habenula nucleus: an *in vitro* study. *J.Neurophysiol.* **59**, 212-225

Information Processing by

Oscillating Neurons

C.L.T. Mannion[*]

J.G. Taylor[+]

[*] Department of Electrical Engineering, University of
Surrey, Guildford, Surrey.

[+] Department of Mathematics, King's College, Strand,
London, WC2R 2LS

Abstract

This paper reviews the methods by which coupled oscillating
neurons may be used by various brain modules to process
information.   In particular the binding and separation
problems are discussed in the oscillatory framework, and
tentative models proposed for how these problems may be
solved.

1. Introduction

There are numerous regions in the brains of higher animals where coupled oscillations have been observed. A partial list is:

- Olfactory cortex and bulb ([1])

- Primary visual cortex ([2])

- Auditory cortex ([3])

- Hippocampus ([4])

- Thalamus ([5])

- Inferior Olive [6]

The range of oscillations can be from below 0.1 Hz up to the gamma frequency range (30-80 Hz). The discovery of correlation between oscillations in different single cells in visual cortex in 1987, [2] has, in particular, aroused interest in the neural network community as a new paradigm for artificial neural net computation. The difficult problems of object fusion and of simultaneous object separation have been conjectured as being solvable by coupled neuronal oscillations [2]. These ideas involve respectively the binding of the output from feature detector cells activated by a common object against a background, and the separation from each other of these for a set of different objects against some background. The use of ensemble coding in terms of averaged activity has not proved satisfactory to solve either these problems [7]; various attempts have been made recently to use coupled oscillators to achieve feature binding.

It has been proposed that, beyond low level analysis of inputs, coupled oscillations may also be involved in attentional processes [8], and in category and correlational learning. Even higher level processing has been suggested as involving coupled oscillatory modules [9]. This has gained interesting support from recent results on coherence in magnetic brain wave scans filtered at 40 Hz [10]. It is possible that the coupled oscillatory approach to neural network information processing will have more success than that based on ensemble coding by average firing rate. However the new avenue will require careful analysis of temporal correlations, even down to short time periods of about 5msec. This means that temporal properties of single neurons may need to be incorporated at a suitably sophisticated level in order to utilise the properties of such oscillatory activity to the full.

The purpose of this paper is to review the above questions in the light of the known experimental data, and in particular on that associated with the recent magnetic field measurements of [10]. In order to pursue this, we will have to consider the nature of models of single neurons which are appropriate for

the study, before these can be applied to modelling
information processing.   The full set of topics we wish to
consider are:

(a) neuronal models
(b) nature of neuronal interconnections
(c) nature of neuronal plasticity
(d) nature of input information
(e) nature of the processing to be performed

These will be discussed in more detail in the paper.

2. The Data

It is only possible to present here a brief survey of the
experimental literature on oscillations in the brains of
higher animals.

2.1 Olfactory System

There has been an enormous amount of experimental results
obtained on oscillations in both olfactory bulb and cortex
[1], [11].   The electroencephalographic (EEG) records show
sinusoidal bursts of gamma oscillations (35–85Hz) on
inspiration of an odor, with a spatial pattern which varies
according to the specific odor being inhaled.   The manner in
which cross-correlation and phase dispersion vary between bulb
and cortex has been studied by Bressler [12].   He found that
significant variation in both correlation and phase dispersion
occurred at different positions for pairs of sites recorded in
the olfactory bulb and cortex.   Mean bulbocortical correlation
was found to be inversely related to phase dispersion and that
the interdependence between bulb and cortex is spatially non-
uniform.   The accessory olfactory bulb (AOB) is of interest in
being simpler (having fewer cell types) and being involved
with simpler learning behaviours.   A particularly interesting
case is that involved with abortion in a female rat by non-
stud male odors during a critical period [13].   This has
recently been modelled by coupled oscillators [14] but there
is presently no data on the modification of the oscillatory
activity related to this learning activity.

2.2 Primary Visual Cortex

The original data of [2] that nearby single cells in visual
cortex have correlated   oscillations over several cycles,
have been confirmed and expanded by the original researchers
[15] and by others [16].   These have shown that there is
inter-columnar synchronisation of oscillatory activity between
cells with similar orientation specificity, even for cells as
far apart as 7mm, and in different cortical areas [16].
Synchronisation was also noted for cells with the same
receptive field but different orientation specificity.

The input stimulus dependence of these correlations have been studied carefully [15]. One of the important results [15] is that there is roughly stimulus intensity independence of the frequency of the oscillations (once the correlation was turned on by the inputs). For example there was only a small change in frequency (well within the experimental error) for a wide range of illuminated slit velocities or lengths, although the amplitude of the oscillatory component of the local field potential varied considerably under such conditions.

It was claimed [15] that there is no concomitant oscillation in LGN. The relevance of the lack of thalamic oscillation in modelling studies may be questioned, especially since the above results were obtained in the case of anaesthetised animals. The results of the existence of synchronised cortical oscillations are clearly relevant to the problem of stimulus binding. The possibility of coupled thalamic oscillations will be left open, especially following the results of [10].

2.3 Auditory Cortex

The existence of an event-related potential (ERP) in the gamma range in auditory processing has been observed by various groups. This is made especially evident if clicks are presented to the subject at 40 Hz [3], but 40 Hz oscillations are observed over several cycles if the clicks are presented at other frequencies [17]. It has been suggested that this 40 Hz response may be used as a characteristic of the awake state, and that the absence of the 40 Hz peak in the ERP power spectrum may be used as a signature of the anaesthetised state when surgery is being performed [17] [18]. There are various negative and positive response peaks in the ERP spectrum, but the only one observed to give a resonance appears to be that at about 40 Hz.

2.4 Hippocampus

The theta rhythm at 3-12 Hz has been well documented as being present in the various hippocampal cell fields during certain forms of activity [19] [20]. Theta has been proved to be generated, at least in part, by the septal cholinergic input which is fed to hippocampus, onto inhibitory interneurons. These then combine with the excitatory pyramidal neurons to produce the theta rhythm. It has been suggested as being used to give a global clock for information processing in the various parts of hippocampus [21], although this has yet to be shown valid. Phase locking between entorhinal cortex and hippocampal cells at theta frequency has also been observed. There are two other patterns of activity in hippocampus, sharp waves with frequencies up to 800 Hz and desynchronised activity.

2.5 Thalamus

There are various rhythmic activities which may be observed in thalamus, of which the most important seems to be spindling, bursts of activity at about 7-14Hz which last about 1 to 2 seconds and recur periodically with a slow rhythm of about 0.1 - 0.2 Hz. This spindling is observed during natural sleep and under barbiturate anaesthesia [5]. There is very important involvement of thalamic and thalamic reticular nucleus cell activity in all ongoing cortical information processing, beyond the 'gating activity of thalamic relay systems'. Some aspects of this will be discussed later.

3. Neuronal Models and Interconnections.

There are two different influences at work determining the dynamical activities of neural networks. The first of these is the intrinsic properties of single cells. These have been discussed in detail, for example, by Llinas and co-workers [22], where the origin of two intrinsic frequencies at 6Hz and 12 Hz for thalamic cells were indicated. The two different responses arose from the presence of a Ca^{++} activated current which is disinhibited by the presence of hyper polarization for a long enough period, as compared to the usual fast Na channel. Intrinsically oscillating cells at about 40 Hz have also been noted as existing in layer 5 in cortex [23]. These cells are thought to be inhibitory interneurons, and have a sub-threshold membrane potential oscillation at around 40 Hz. There are also cortical cells with a frequency response proportional to input stimulus, but these may not necessarily play an important role in the 40 Hz generation. They could be essential in all other aspects of information processing using oscillatory neurons.

The other aspect of neural structures important in the generation of oscillatory cell output is the internal network architecture. Inhibitory-excitatory feedback pairs of neurons form a basic circuit here. The activity of one neuron of the pair excites the other until it is so strongly excited that its feedback inhibition finally turns off the first. This then dies away, so turning off its inhibitory partner, which then releases the first cell from inhibition. This then starts the cycle again. A typical system of equations describing this activity is of the form, for leaky integrator neurons,

$$\dot{x} = -x - f(y) + I \qquad (1a)$$

$$\dot{y} = -y + f(x) \qquad (1b)$$

where I is the input, the time constants have been set to 1 in (1) for both neurons, and f the usual sigmoidal type of function giving output rate in terms of membrane potential x or y. The purely damped oscillatory behaviour for the linear case $f(y) = ay$ has transfer function in the Fourier plane $[1+ a - w^2 - 2iw]^{-1}$ with response frequency $2\pi/a$ and damped lifetime 1. Similar oscillatory properties arise on linearising f around a value at which x = y = 0, and is the basis of the analysis of Freeman and co-workers [11], Baird [24] and Li and Hopfield [25] on odor coding in olfactory bulb and cortex. In that case the leaky integrator system is extended by regarding x and y as column vectors describing the activity of the excitatory and inhibitory neurons. Connection matrices $\underline{G}$ and $\underline{H}$ are then introduced, so that equation (1) becomes the coupled system of equations

$$\dot{\underline{x}} = \underline{-x} - \underline{G} \underline{f} (\underline{Y}) + \underline{I} \qquad (2a)$$

$$\dot{\underline{y}} = \underline{-y} + \underline{H} \underline{f} (\underline{x}) \qquad (2b)$$

and may be extended further to coupled modules of E-I feedback systems. The general nature of the dynamics of such systems has been discussed in some detail in [24], and simulations performed by [26].

Modelling of intrinsic neuronal oscillators may be attempted at the most naive level by regarding them as coupled S^1's [27], [28]. A considerable body of knowledge exists about the nature of responses of a system of coupled S^1's to external inputs, in terms of phase structure in the Ising model, although transient behaviour is not well studied. the equations for a set of such coupled oscillators, in terms of phases f_i, is

$$\dot{f}_i = I_i + \Sigma_j a_{ij} H(f_j - f_i)$$

where I_i are external inputs, a_{ij} are connection weights, and this a periodic function.

One may add non-linearity in two ways. One is to incorporate properly the various ionic channels, with their time constants, into the single neuron equations. This may be done with the full Hodgkin-Huxley approach, or use the more tractable Fitzhugh-Nagumo model. This has been discussed by us in this context elsewhere [28] and we will not dwell on it here (see also the analysis of the piecewise linearised version in [29]).

The second way of incorporating non-linearity in a non-trivial fashion is to add delayed self-inhibition or self-excitation terms to the activation term in a binary decision, or more correctly a leaky integrator neuron model. This may be achieved in (1) by adding the terms proportional to $\int^t dt^1 \cdot (\exp[-(t-t^1)/a]) \cdot x(t^1)dt^1$ and $\int^t \cdot dt^1(\exp[(t-t^1)/b]) \cdot y(t^1)dt^1$ to the right-hand side of (1a) and (1b) respectively. Depending on their signs, these terms will modify the threshold for activation in an activity-dependent manner either excitatorily or inhibitorily. Non-linear functions of x or y may also be used as the integrands of these threshold variations, to attempt to mimic better the actual non-linearities of these neurons.

Finally we note that if the extensive presence of feedforward inhibition is incorporated, it may be a good enough approximation to drop the leakage terms on the left-hand sides of (1) and (2), so leading to a net of purely binary decision neurons. The frequency dependence of a single one of these (also provided with feedback inhibition) has been studied in [30]; the case of a coupled net of such neurons can be analysed by an extension of this approach.

We will not discuss neuronal plasticity separately here, but refer to [28] where this is considered in the context of coupled neuronal oscillators.

4. Information Processing by Coupled Oscillators.

In this section we wish to discuss the two basic problems facing early processing which were raised in the introduction:

(a) fusion or binding

(b) separation of bound objects (so that more than one object may be processed in parallel across coupled modules);

Such segmentation may be at the basis of the magic number 7 ± 2 of object detectable in short presentations to human subjects.

It is important at this point to note the results of Llinas and Ribary [10] on the existence of what appears to be a global 40-Hz synchronization of brain magnetic field activity, in which there is a 12.5 msec phase difference between front and back of cortex; the frontal pole is in the lead. There is also evidence [10] that there is concomitant synchronised 40-Hz brain stem activity which is about 3msec ahead of the cortical activity. This supports the notion that there is an excitatory-inhibitory feedback loop between thalamus and cortex which is strongly excited when information processing is occurring. Such a loop must involve the reticular nucleus of the thalamus, whose inhibitory feedbacks to thalamus and excitatory inputs from thalamocortical and corticlthaoamic axons are well documented. There is preliminary evidence for such linkage in the second paper in [10].

Putative oscillators in thalamus and cortex (mentioned in sub-section 2.2) are excited by inputs. For a given input there are required to be excitatory connections between cortical pyramidal cells which achieve phase locking. Simulation of small nets [26], [27], [28] with such interconnections has shown that this is possible. The creation of the lateral connections would seem achievable by a suitable learning rule, and has already been considered in [28]. Thalamic (and Reticular) activity may be of importance to achieve tight phase synchronisation.

The coding of a single object would thus appear to be achieved by a phase syncronised ensemble of cells. Can another ensemble of cells code almost simultaneously for another object? It might do so by using a different frequency. In order to do this by the mechanism of the previous paragraph it would be necessary to observe oscillator cells in cortex with a range of stimulus - independent frequencies; these have not been observed, and the most relevant results is the quoted value of 42±7Hz for stimulus driven oscillations for 20 cells by Llinas [23]. On the basis of this evidence, we feel it worth while to explore the model in which objects are not coded by frequency but by phase and amplitude. In other words all information processing in our model is supposed to occur by means of synchronised 40 Hz oscillations.

There is psychological evidence of the existence of 25 msec as a natural unit or quantum of time in the central nervous system. Thus Poppel [31] reports on the temporal order threshold (the minimum time needed to perceive the order in time of two events) for different mobilities as 20 msec or so, and that exhaustive scanning can occur at an approximate speed of 30 msec per item. If two random patterns that together form a word are shown for 6 msec each, the word will be perfectly recognised with a 25 msec temporal interval between the patterns, much less often with a 50 msec interval [32]. Work on reaction time also supports the notion that 30 msec is a basic temporal processing unit. Thus the histograms of choice reaction time are in most instances multimodal with a 30-40 msec interval between adjacent modes [33]. Moreover careful tests show that the oscillations are stimulus driven and specifically coupled to the onset of the stimulus. This agrees with the use in our model of the cortical layer 5 inhibitory interneurons of Llinas and colleagues [23] which oscillate at 40 Hz in such a stimulus driven manner.

In order to detect, say, 6 objects, it would be possible to allow the synchronised neural ensemble for each object to have activity for at most 6 msec out of the 25 msec available for a whole period at 40 Hz. This leads us to the problem of phase separation: how to divide up the synchronised oscillatory activity so that these phase separations between different sub-ensembles can be achieved. That such a separation can also occur across modules appears evident from the Llinas and Ribary result [10]; how is it to be obtained even in a single module if at all?

An alternative approach is to consider that each of the 6 objects is characterised by one or two cycles of the 40 Hz oscillations. If 300 msec time is allowed as that used to process a visual scene between saccades, then 6 objects could be processed serially, one every 50 msecs or so.

We call this latter model the serial processing model (SPM), as compared to the earlier one which we term the parallel processing model (PPM). An interesting model of this sort has been developed in [34], although is rather limited in its response to input patterns since the connection weights were determined by a Hebbian rule. Only a few of the many possible input patterns can be stored in this way; a more flexible system is needed. One could also consider exactly parallel processing, in which several sub-ensembles of neural activities are exactly in phase; this does not help solve the separation problem at all.

The SPM would, on the face of it, allow more laxity in temporal precision compared to the PPM, where even a few milliseconds of temporal delays in signal passing could cause confusion between two objects. The SPM model also has its difficulties, the most important of these being how the information of the later of the 6 or so objects being processed in the 300 msec time block is stored until it achieves its synchronisation. It would seem to be more seriously vulnerable to degradation than the earlier ones, which does not happen so easily in the PPM model. These models of information processing lead to different predictions for psychological and MEG tests, which will be discussed elsewhere.

In order to determine the viability of the SPM and PPM models we have constructed a simple analytic form of the first of these models, and simulated them.

5. A Simple Serial Processing Model

The model is based on a single layered cortex and a thalamus. In the cortex there is a set of pyramidal cells with excitation on them from the thalamic relay cells and inhibition from inhibitory cortical interneurons, acting as intrinsic oscillators. Both sets of cortical cells are driven by the thalamic cells, which themselves are driven by input. The complexity of the reticular layer is neglected, although an assumption is made that either by adaptation or by feedback inhibition in the glomeruli, there is a self-inhibition term on each of the thalamic relay cells which builds up as the activity proceeds (similar to that used in [34]).

For simplicity only average firing rates are considered, so that neurons are modelled by means of the equations of type (1). Let Y_1, Y_2, Y_3 denote cell potentials for the inhibitory cortical neuron, the cortical pyramidal cell and the thalamic relay cell. Then the coupled equations for a primitive unit are:

$$\dot{Y}_1 = -s^{-1}Y_1 + as^{-1} Y(Y_3) \cos wt \qquad\qquad (3a)$$

$$\dot{Y}_2 = -s^{-1}Y_2 + b^{-1} Y(Y_3) - cs^{-1}Y(Y_1) \qquad\qquad (3b)$$

$$\dot{Y}_3 = -s^{-1} Y_3 + s^{-1} I - s^{-1}d \int^t (\exp[-(t-t^1)/e]) Y(Y_2(t^1))dt^1 \qquad (3c)$$

In (3) the coupling coefficients are all normalised by the same constant, Y is the unit step function (in simulations smoothed to a sigmoid function with temperature of 0.1, s is the membrane time constant of all neurons, e is the time constant of the cortical thalamic feedback inhibition and t_1 is the time when the thalamic neuron Y_3 has activity above a certain threshold. For suitably small s and e the equations (3) can be solved analytically. The solutions for a constant input current I in (3c) have the Y_1 cell oscillating for a time of O(e), with the Y_2 cell also oscillating, but in anti-phase. The inhibitory feedback from the Y_2 cell onto the Y_3

cell builds up, with a time constant of e, till ultimately output from that cell ceases to arrive on the Y_1 or Y_2 cells. The oscillations then cease.

In figure 1 is shown the activities of the three cells of a single vertical unit when the self-inhibitory term in (3c) is dropped and a constant input is applied. The oscillations of the pyramidal cell are clearly in anti-phase with that of the inhibitory interneuron, whilst that of the thalamic cell rapidly ramps up to a constant value. The values of the parameters are given in the figure legend. The effect of the self-inhibitory term is shown in figure 2, in which the pyramidal cell dies out after about 4 or so oscillations with constant input. Finally, even without lateral coupling two different inputs onto a pair of these units causes oscillation initially for the low input region to be turned off very rapidly, leaving that for the high input region to give one or two periods. This then turns itself off, and the low input oscillations turns back on again as seen in figure 3. Use of a modification of threshold brought about by the spread of activation in the reticular layer would be expected to implement this sequential mode of oscillatory ensembles, and is presently being investigated.

Lateral coupling may be included in all of the equations (3), with the neurons arranged on three two-dimensional sheets. Using matrix terminology for the coupling coefficients, with $a_{ij}(\underline{r}, \underline{r}^1)$ denoting the weight for the effect of neuron of type j at position $\underline{r}^1$ on neuron of type i at $\underline{r}$ (i,j = 1,2,3), we have:

$$\dot{Y}_1(\underline{r}) = -S{-}1Y1(\underline{r}) + s^{-1} \int d^2\underline{r}^1 \, a_{13}\,(\underline{r},\underline{r}^1) \, Y(Y_3(\underline{r}^1))\,\cos wt \quad (4a)$$

$$\dot{Y}_2(\underline{r}) = -s^{-1}Y_2(\underline{r}) + s^{-1} \int d^2\underline{r}^1 \, a_{2j}(\underline{r},\underline{r}^1)) \quad (4b)$$

$$\dot{Y}_3(\underline{r}) = -s^{-1}Y_3(\underline{r}) - I(\underline{r}) - s^{-1} \int \{\exp[-(t-t^1)k]\} \, a_{32}(\underline{r},\underline{r}^1) \, Y$$
$$(Y_2(\underline{r}^1,t^1))d^2\underline{r}^1 + s^{-1}\int d^2\underline{r}^1 \, a_{33}(\underline{r},\underline{r}^1) \, Y\,(Y_3(\underline{r}^1)) \quad (4c)$$

It is reasonable to choose for the coupling coefficients $a_{ij}(\underline{r},\underline{r}^1)$ a sum of Gaussians

$$a_{ij}(\underline{r},\underline{r}^1) = m_{ijk}\exp(-|\underline{r}-\underline{r}^1|^2/2d_{ijk}^{-2}) \quad (5)$$

We take the self-coupling term a_{33} to have Mexican hat shape, so as to lead to a competitive net, whilst a_{22} can be taken as having a similar form, although with perhaps a shorter repulsive part (corresponding to inhibition at a distance of half a hypercolumn and a longer range attractive part). a_{21} is taken to be a narrow positive Gaussian, as are, a_{23}, a_{32}, a_{21} and a_{13} (corresponding to preserving topographic mapping). This more extensive model is also being considered presently. It may also be analysed analytically by Fourier techniques. Initial results indicate rather similar response as that of the previous paragraphs.

We conclude that equations (3) give a very simple instantiation of the serial processing mode of primary cortical activation. The model has great flexibility in

handling inputs, not depending on learned connection weights.
However short-term potentiation could be used to sharpen up
the processing; long-term potentiation, along the lines of
[34], would only seem to make the processing rather
restricted.

6. Later Processing

The information arising from synchronised activity in primary
cortical cells at different positions must be combined
together to be used by later modules. In particular the use
of semantic analysis of the inputs must occur, as well as the
relation of the resulting higher level activity to stored
memories for recognition processes. Storage of such activity
may also be brought about. Attention may be involved here,
and will be included in our analysis.

The results of [10] indicate that such processing may well
occur in a globally well-defined sequence. In attentive
processing frontal lobe 40-Hz activity spreads back caudally
to associative then to primary sensory regions, then ventrally
to temporal lobe and finally back to frontal lobe. Such
sequencing fits in very well with modern models of the frontal
lobes [35] where there is planning, and inhibition of
distractors to ongoing plans. The strong connections from
frontal to parietal lobe is especially relevant to attention
[36], and may be modelled in terms of a phase differencing
system in the parietal attentional centre. The comparison of
phases may be set up by excitatory feedback to upper cortical
layers as compared to feed forward excitation which also
excites the intrinsically oscillatory inhibitory interneurons
of type (3a). Assuming the frontal feedback is already
oscillatory, the output of the parietal pyramidal cell (3b)
will now be proportional, in a simple case, to the differences
$\cos(at + b) - \cos(at + c)$, where b and c are the phases of
the frontal oscillatory feedback and the feedforward cortico-
cortical from primary or associative cortex (a corresponds to
the 40 Hz frequency, and we have dropped a phase difference
due to rostral lead of the 40Hz). The parietal pyramidal cell
output will be proportional to $\sin(b-c)$. This is maximum for
a quater-wave phase difference, but vanishes for phase
equality. The attentional model has the level of attention
proportional to the output of the parietal cell. If the input
agrees with frontal lobe expectation (b=c) then attention to
this particular input will be zero; attention will disengage
[36] to another input which has non-zero phase difference with
the expectancy wave from frontal lobe.

When attention has moved it may re-engage partly by the same
mechanism. There is evidence [36] that the reticular nucleus
(RN) is also involved in that activity. We have not modelled
RN here but hope to do so elsewhere.

References

[1] W.H. Freeman, Mass Action in the Nervous System, Academic Press, 1975.

[2] C.M. Gray and W. Singer, IBRO. Abstr. Neurosci. lett. suppl. 22, 1301, 1987

[3] R. Galambos, S. Makeig and P.J. Talmarchoff, "A 40-Hz auditory potential recorded from the human scalp", Proc. Nath. Acad Sci USA, 78, 2643-2647, 1981

[4] J. O'Keefe and L. Nadel, The Hippocampus as a Cognitive Map, Osford Univ. Press, 1978

[5] M. Steriade, D.G. Jones and R. Llinas, Thalamic Oscillations and Signalling, John Wiley, 1990

[6] R. Llinas and Y. Yarom, "Oscillatory properties of guinea-pig inferior olivary neurones and their pharmacological modulation" J. Physiol (Lond) 376, 163-182, 1986

[7] W. Singer, "Search for Coherence: A Basic Principle of Cortical Self-Organizations", Conception Neuroscience 1, 1, 1989

[8] F.H.C. Crick and C. Koch, "Towards a Neurobiological Theory of Consciousness", Seminars in Neuroscience (to appear)

[9] D.E. Sheer, "Sensory and Cognitive 40-Hz Event-Related Potentials: Behavioural Correlates, Brain Function and Clinical Application", 339-374 in Brain Dynamics, ed E. Basar and T.H. Bullock, Springer 1989

[10] R. Llinas and U. Ribary "Rostrocaudal Scan in Human Brain: A Global Characteristic of the 40-Hz Response During Sensory Input", in Induced Rythms in the Brain, E. Basar and T. Bullock, eds, Burkhauser, in press

[11] Y. Yao and W.H. Freeman "Model of Biological Pattern Recognition with Spatially Chaotic Dynamics", Neural Networks 3, 153-170, 1990

[12] S.L. Bressler "Relation of Olfactory bulb and cortex", Brain Res. 409, 285-301, 1987

[13] P. Brennan, H. Kaba and E.B. Keverne, "Olfactory Recognition: A Simple Memory System", Science 250, 1223-1226, 1990

[14] E.B. Keverne and J.G. Taylor "Accessory Olfactory Learning", Biol Cyb. 64, 301-305, 1991

[15] C.M. Gray, A.K. Engel, P. Konig and W. Singer "Stimulus Dependent Oscillations in Cat Visual Cortex", Europ. J. Neuroscience 2, 588-607, 607-619, 1990

[16] R. Eckhorn, R. Bauer, W. Jordan, M. Brosch, W. Kruse, M. Munk and H.J. Reitbock, "Coherent Oscillations: A Mechanism of Feature linking in the Visual Cortex?", Biol. Cyb. 60, 121-130, 1988

[17] E. Poppel, "Auditory Evoked Potentials Indicate the loss of Neuronal Oscillations During General Anaesthesia", Naturwiss 74, 542, 1987

[18] J. Kulli and C.Koch "Does anesthesia cause loss of consciousness?", Trends in Neurosciences 14, 6-10, 1991

[19] G. Buzaki, "Generation of Hippocampal EEG Patterns", in The Hippocampus, vol 3, ed Isaacson and K. Pribram, Plenum Press, 1986

[20] M. Stewart and S.E. Fox, "Do septal neurons pace the hippocampal theta rythm?", Trends in Neuroscience, 13, 163-168, 1990

[21] B.L. McNaughton and R.G.M. Morris, "Hippocampal synaptic enhancement and information storage without a distributed memory system", Trends in Neurosciences 10, 408-415, 1987

[22] R.R. Llinas, "The Intrinsic Electrophysiological Properties of Mammalian Neurons: Insights into Central Nervous System Function", Science 242, 1654-1664, 1988

[23] R.R. Llinas, A.A. Grace and Y. Yaron, "In vitro neurons in mammalian cortical layer 4 exhibit intrinsic oscillatory activity in the 10 to 50-Hz frequency range", Proc Nath Acad Sci USA 88, 897-901, 1991

[24] B. Baird "Nonlinear Dynamics of Pattern Formation and Pattern Recognition in the Rabbit Olfactory Bulb", Physica 22D, 150-175, 1986

[25] Z. Li and J.J. Hopfield, "Modelling the Olfactory Bulb and its neural Oscillatory Processing", Biol. Cyb 61, 379-392, 1989

[26] R. Eckhorn, H.J. Reitboeck, R. Dicke, M. Arndt and W. Kruse "Feature linking Across Cortical Maps via Synchronisation", pp101-104 and P. Konig and T.B. Schillen "Segregation of Oscillatory Responses by Conflicting Stimuli", pp 139-142 in Parallel Processing in Neural Systems and Computers, ed R. Eckmiller, G. Hartmann and G Hauske, North Holland 1990

[27] D.M. Kammen, P.J. Holmes and C. Koch, "Origin of Synchronized Oscillations in visual cortex : global feedback versus local coupling" preprint 1990.

[28] C.L.T. Mannion and J.G. Taylor "Coupled Excitable Cells", in <u>Advances in Neural Computation</u>, ed C.L.T. Mannion and J.G. Taylor, Springer (in press).

[29] L.F. Abbott, J. Phys A. Math Gen. <u>23</u>, 3835, 1990

[30] E.R. Cainniello, "A study of neuronic equations", pp187-199 in <u>New Developments in Neural Computing</u>, ed J.G. Taylor adn C.L.T. Mannion, Adam Hilger, 1989

[31] E. Poppel, "Time Perception" pp1215-6 in Encyclopedia of Neuroscience, Birhauser, 1987

[32] C.W. Eriksen and J.F.Collins "Sensory traces versus the psychological moment in the temporal organization of form", J.exp Psychol. <u>77</u>, 376-382, 1968

[33] E. Poppel, Naturwissenschaften <u>55</u>, 449, 1968

[34] W. Schneider and Ch. von der Malsburg "A Neural Cocktail-party Processor", Biol. Cyb. <u>54</u>, 29-40, 1986

[35] D.T. Stuss and D.F. Benson <u>The Frontal Lobes</u>, Raven Press 1986

[36] M.I. Posner and S.E. Petersen "The Attention System of the Human Brain", Ann. Rev. Neurosci <u>13</u>, 25-42, 1990.

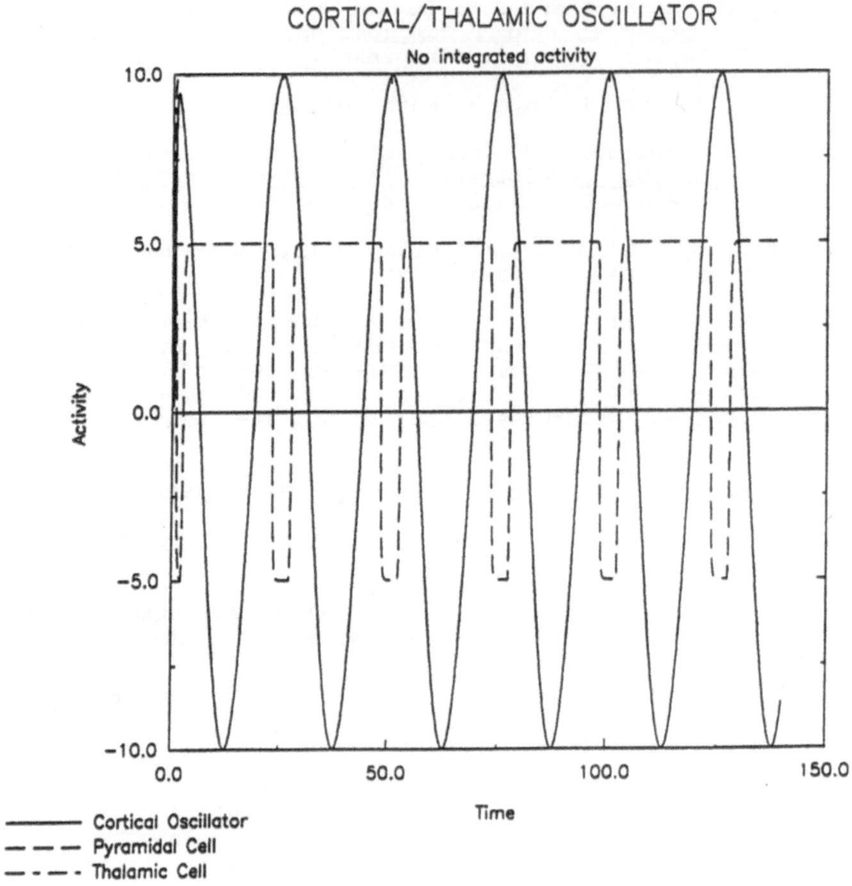

CORTICAL/THALAMIC OSCILLATOR

FIGURE 1.   Response of the coupled cortical-thalamic system of equation (3);
the figure is explained in the text.

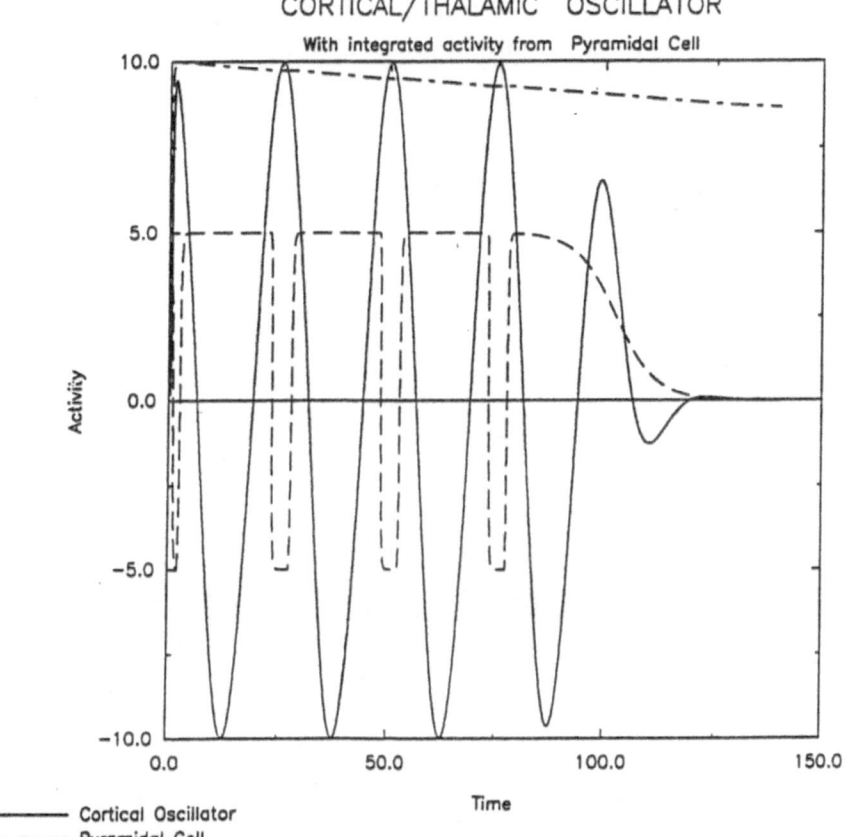

CORTICAL/THALAMIC OSCILLATOR
With integrated activity from Pyramidal Cell

Cortical Oscillator
Pyramidal Cell
Thalamic Cell

FIGURE 2.   Similar to fig, but now with integrated self-inhibition in equation (3c)

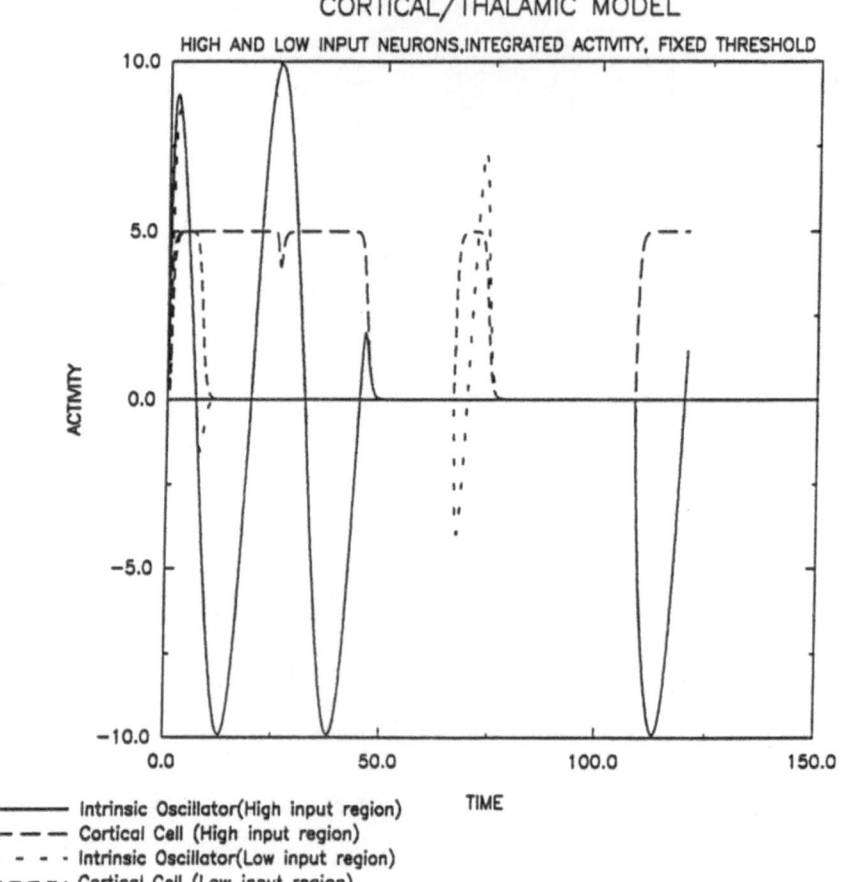

CORTICAL/THALAMIC MODEL

HIGH AND LOW INPUT NEURONS,INTEGRATED ACTIVITY, FIXED THRESHOLD

——————— Intrinsic Oscillator(High input region)
— — — Cortical Cell (High input region)
- - - - · Intrinsic Oscillator(Low input region)
— — — — · Cortical Cell (Low input region)

FIGURE 3.   Simulation of the serial processing model

GAMMA OSCILLATIONS, ASSOCIATION AND CONSCIOUSNESS

Rodney M.J. Cotterill and Claus Nielsen
Division of Molecular Biophysics
The Technical University of Denmark
Building 307, DK-2800 Lyngby, Denmark

ABSTRACT

A model which incorporates the recently-explored connections between the various layers within the striate cortex (V1), and also the projections (forward and reverse) between that area and an extra-striate area (ES) has been shown to support oscillations in the gamma frequency range, these occurring as an inter-area phenomenon rather than one confined to a given area. The dynamical behaviour of the model suggests how the special properties of NMDA receptors might be exploited, and incorporation of this possibility produced a mechanism whereby inputs from two different sensory modalities could be associatively linked. The characteristics of the observed oscillations also provided direct explanations for three commonly identified features of consciousness, namely very short memory, the discrimination of coherent components in a given sensory input, and the binding of different sensory inputs that are causally related.

INTRODUCTION

The recent discovery[1-3] of oscillations in the gamma range, in the striate cortices of adult cats, due to the movement of suitably-oriented bars across their visual fields, has provoked considerable excitement in the neurophysiological fraternity. There is a growing feeling that a proper explanation of these oscillations is going to shed important light on the way in which the visual system processes incoming information[4]. Indeed, there are many who feel that a full appreciation of the significance of these new observations is even going to elucidate the mechanism of consciousness itself[5,6].

In this paper, we will present the results of an exploration of the properties of a cortical model that incorporates some of the latest information on the way in which the various parts of the visual system are interconnected. Such information is now available not only for the various layers of the striate cortex (V1)[7-9] but also for the pathways between V1 and what are usually referred to as the higher areas. In particular, information is now available on the routes taken by the projections that connect V1 to extrastriate areas (ES)[10,11]. This newly-won data has transpired to be of paramount importance to the understanding of the functioning of the primary visual system, as will be described in detail.

But the model that will be described here admittedly suffered from one serious limitation: the cortical circuitry believed to underlie the brain's ability to detect motion in its visual field was not incorporated. We have plans of doing this at a later stage of our investigations, but our studies have yet to reach this obviously desirable degree of sophistication. Despite this shortcoming, we feel that our model displays behavioural characteristics that would be more general than those merely associated with the actual faculty addressed, namely the detection of static coherent features in the visual system; we suspect that the phenomena displayed by our limited model would be quite general in the visual system, and that they, or something rather like them, would also been seen in those parts of the system that specifically handle the detection of motion. We believe that this can be stated with some confidence because the anatomical features important to our model's behaviour are observed to be quite widespread in the cerebral cortex.

We will first give a description of the relevant anatomical details and then move on to the way in which these were built into the model. This will be followed by an account of the model's basic characteristics, including the important observations on its oscillatory behaviour, and we will then move on to a description of the association mechanism that follows when these oscillations are combined with a plausible incorporation of NMDA-mediated learning. Finally, we discuss how the fundamental properties of the model offer relatively simple and straightforward explanations of three of the features commonly believed to be associated with consciousness.

ANATOMICAL BACKGROUND

The anatomist now has at his disposal a considerable battery of techniques which permit the charting of the various routes taken by the processes of neurons, as they link up the different parts of the cerebral cortex, and also the various regions of the latter to the equally important subcortical parts of the brain. In particular, the early parts of the visual system have now been intensely studied, and rather detailed circuit diagrams are available which reveal the manner in which the various cortical layers are interlinked. In some cases, optical microscopy has been augmented by electron microscopy at the sub-synapse level, and this has provided supplementary information as to which of these inter-layer linkages are excitatory and which are inhibitory.

Just as importantly, the anatomists have now been able to demonstrate how some of the cortical areas in the visual system are linked together. It has transpired that these links generally run in both directions, which is to say that a given area lying early in the hierarchy sends projections to other areas lying "farther up" and that it also receives similar (reverse) projections from that area. This does not mean, however, that such projections are precisely reciprocal. On the contrary, this is not the case in two important respects. The layers which send and receive such projections are not the same for the forward and reverse directions, and the projections do not in general always involve just pairs of cells. This point will be made clear in a later section of this paper, where it will be shown that the lack of strict reciprocity has important ramifications for the visual system's powers of discrimination.

Armed with this newly-acquired circuitry information, one is now able to construct a diagram which displays all the relevant linkages between the layers in a given area, as well as the projections that bind the various areas in the early visual pathway. Such a circuit diagram is shown in Fig. 1, and it was this which provided the basis for the simulations that will be described in this paper.

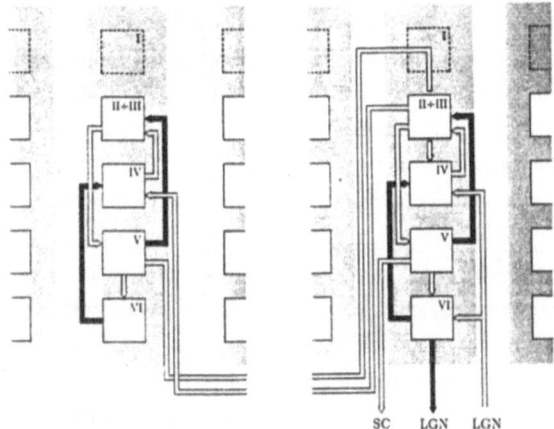

Figure 1. The connections between the different layers of a given cortical area have been revealed by recent anatomical investigations, as have the forward and reverse projections via which pairs of interacting cortical areas are mutually linked. Both types of connection are shown in this diagram, which were incorporated in the computer simulations described in the present paper. The unshaded rectangles denote the individual neuronal assemblies.

The vertical shaded rectangles shown in Fig. 1 denote minicolumns (about 30 μm wide), each of which is a grouping of a certain number of neurons. Counts of cell numbers in columns of cortical tissue[12] give remarkably constant numbers of 110 ± 10 cells in many cortical areas and species, exept in primate visual cortex where the number is about 270. The long axis of such a minicolumn lies normal to the plane of the cerebral cortex, and it spans the entire depth of the cortex from pia to the white matter. It thus traverses all six of the cortical layers, the second and third of which are now usually, in the visual processing areas, taken to perform as a single composite layer, designated II+III. Layer I contains very few cell bodies, and it is safe to assume that this outer cortical region is mainly concerned with the formation of synaptic connections. Such minicolumns formed by interconnected cells are not defined by anatomical boundaries, but rather form their homogeneous activities caused by response to specific stimuli[12-15]. It is thus realistic to delimit groups of highly interconnected cells sharing the same set of inputs and outputs.

Of particular significance is the fact that the reverse projections[10,11] from a "higher" cortical area are generally observed to pass from the white matter successively through layers VI, V, IV and II+III, without establishing synaptic contacts with the neurons in any of these layers. Instead, they reach Layer I, there to form synapses onto the elongated dendrites of neurons whose somatic regions are actually in layer IV[16].

It is important to emphasize that Fig. 1 can be misleading in one important respect. A too literal interpretation of the layout there depicted could lead one to believe that the forward and reverse projections form neatly reciprocal pairs. This is not the case, for two important reasons. For a start, the establishment of such a reciprocal connection would imply an accuracy of the mapping from one cortical area to another that would be quite beyond the capabilities of the underlying tissue differentiation; for this to have been possible, each minicolumn would have had to be endowed with its own brand of nerve growth factor, which is certainly not the case. One must assume, therefore, that most of the closed loops of projections implied by Fig. 1 would in fact not be closed at all.

The other factor which militates against such a simple and tidy picture is still more important, and it is the result of what we will refer to as "vergence". This word does not appear in the Oxford Dictionary, but we feel that it deserves to be coined, because it neatly expresses the fact that situations often simultaneously involve both divergence and convergence: The word vergence refers collectively to both these attributes, and it conveys a connotation that the word vergency (which *is* to be found in the dictionary) does not. A single neuron in a given cortical layer sends diverging axon collaterals to many neurons in an ensuing layer, while a given neuron in that second layer receives converging connections from many neurons in the first layer: Both the divergence and the convergence stem from one and the same fact, namely that each neuron makes synaptic connections with many other neurons. Use of a single word, albeit a neologism, is thus apposite. The fact of vergence will be seen to be of paramount importance when we later contemplate some of the implications of our model's behaviour.

SIMULATION MODEL

With a single minicolumn containing about three hundred neurons[12], and assuming that these are fairly uniformly distributed between four effective layers, namely II+III, IV, V and VI, there will be about 60–70 neurons in each layer, in a given minicolumn. In fact this is not excatly true. Due to the variation in the relative neuronal densitiy of the layers the number of neurons in each layer-group may vary from the above stated.

These will, moreover, be divided between excitatory(E) and inhibitory(I) sub-species, and we will further assume that this division puts approximately equal numbers in those two categories. A further assumption, which appears justified by the anatomical data, is that the Es excite each other, that they also receive excitation from outside their layer, and finally that they receive inhibition from the Is. The Is also receive activation from the Es and from the outside, but they do not appear to inhibit each other.

One could set up a set of simultaneous differential equations to follow the course of events when such a composite assembly of Es and Is experiences an injection of activity from the outside. But a meaningful interpretation of such a continuum description would be difficult to arrive at because it would be essential to take into account several essentially discontinuous features, such as the existence of an activity threshold for each neuron, the pulsed emission of action potentials when this threshold is exceeded, and the intervention of a refractory period between the generation of two successive pulses. And considering the relevant time scales for the gamma-oscillations (i.e. 25 ms), a far more satisfactory approach is to follow the temporal evolution of the activity levels in the (interconnected) E and I sub-populations by a digital representation.

The input data necessary for such a simulation are the characteristic times for the electrotonic responses of the dendrites, the decay time for the potential at the somatic region, and the duration of the refractory period. In our simulations, these were taken to be 4 ms, 15 ms and 8 ms[17], respectively. It may be noted that our incorporation of a decay in the somatic potential follows what is frequently assumed in such simulations, the modelled neuron often being referred to as a leaky integrator. In the simulations, we found it expedient to incorporate the dendritic and decay time constants into a single temporal profile, the course through which was triggered (for a given axon-synapse-dendrite combination) at the epoch at which the somatic potential was determined to have exceeded the threshold value.

RESULTS

The first thing to be investigated in our study was naturally a single neuronal assembly, with its E and I sub-populations. The temporal evolution of the activities in such a group of neurons was found to be as indicated in Fig. 2, and it is to be noted that the activity level of the Es rises more rapidly than that of the Is because of the mutual excitation

120

in the E sub-population. Ultimately, however, the activity level of the Is predominates, because of their suppression of the Es. These findings is in good accordance with empirical results obtained by Douglas et al. [18].

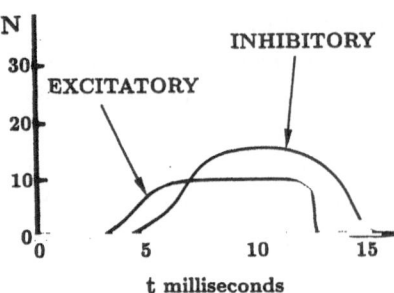

Figure 2. The temporal evolution of the activity levels of the E and I sub-populations in a neuronal assembly comprising sixty neurons, equally divided between the two neuronal types. The essentials of the exhibited behaviour were found to be independent of the actual distribution between the two sub-populations within a factor of two to either side.

Having thus established the essentials of the behaviour of a single neuronal assembly, we then turned to a simulation of the two-area multiple-layer configuration depicted schematically in Fig. 1, each neuronal assembly being treated on an equal footing with all the others. We initially injected activity into layer IV of one of the areas, designated ALPHA, this reflecting the fact that it is this layer in V1 which receives input from the dorsal lateral geniculate nucleus of the thalamus. We then followed the course of the activity patterns in the various layers in the two areas, this being observed to pass first to layer II+III in ALPHA, and thence to layer IV in BETA, to layer II+III in BETA, on to layer V in BETA, and finally back to layer IV in ALPHA. The round trip time for the wave of activity to pass around this closed loop was approximately 21 ms. This corresponds to a frequency of about 50 Hz, which is in good agreement with the reported frequency for gamma oscillations. This temporal course is shown in Fig. 3

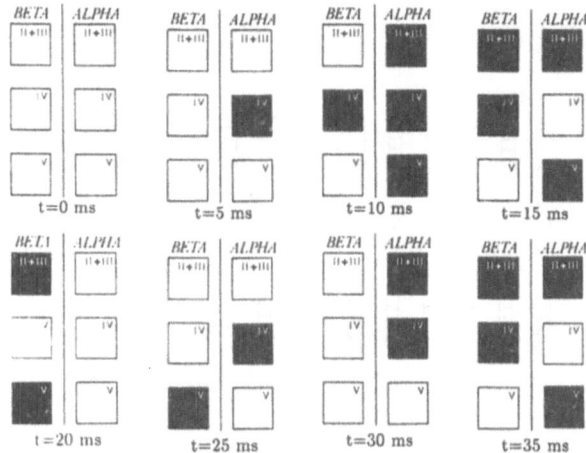

Figure 3. The temporal evolution of the activity levels in the various neuronal assemblies in each of two interconnected cortical areas designated ALPHA and BETA. The white rectangles indicate lack of activity, while the black rectangles indicate that the corresponding assemblies have reached the saturation activity level shown in Fig. 2. The situations are shown at epochs spaced at 5 ms intervals, and comparison of the situations at 15 ms and 35 ms reveal that the periodicity is close to 20 ms.

It is worth reiterating that Fig. 3 could be misleading, just as was the case for Fig. 1, because neither of these illustrations depicts the vergence that was emphasized as being important in a preceding section of this paper. And it is through such vergence that the two-area system acquires another important faculty, namely that of association. A more realistic impression of the situation can be obtained from Fig. 4, which shows the two layers from the side. For convenience, however, they have been shown as if they lay parallel to one another, even though they in fact form contiguous areas of the same continuous cortical sheet. These two arrangements, together with the (white-matter) connections that link them together are, of course, topologically equivalent. For the sake of clarity, only a few of the many synaptic connections are shown, for both the forward and reverse projections, and the important thing to be noted is that a given forward projection between two minicolumns, one in each of the two cortical areas, is not in general matched by a counter-running reverse projection between the same two minicolumns.

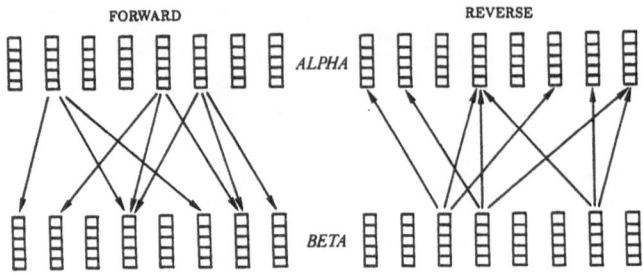

Figure 4. The pattern of forward and reverse projections between two successive cortical areas, shown here as if they lay parallel to each other, rather than as two contiguous areas of the same continuous cortical sheet, these two arrangements being topologically equivalent. It is to be noted that the forward and reverse projections are not merely reciprocals of each other, and that the closed loop implied by Fig. 1 will be a minority feature of the actual situation.

The closed inter-area loop that appears to be implied by Fig. 1 will thus not be a dominant feature of the actual situation, and the transmission of signals back and forth between the two cortical areas can take place perfectly well even if there are no such closed loops whatsoever. And far from being an inadequacy of the system, this distribution of signal transmission is the secret of the system's discriminative powers, for in default of closed loops, with their relatively weak demands on accurate timing of individual impulses, the distributed transmission of signals requires coherence between the various contributing signals. In effect, this means that the only components of an input signal that can contribute to what is passed between the cortical areas will be those that are mutually within step, to within very few milliseconds.

This means that only the coherent parts of an input pattern presented to ALPHA, in the present case, will successfully make the (distributed) round trip via BETA and back to ALPHA. Fig. 5 shows this process occurring, in a simulation in which each of 64 modelled minicolumns in ALPHA engaged in 32 synaptic contacts with minicolumns in BETA, while each of the 64 modelled minicolumns in BETA established a similar number of synapses with minicolumns in ALPHA

As was mentioned earlier, the pattern of forward and reverse projections seen between the striate cortex and area V2 is repeated for the interactions between higher visual

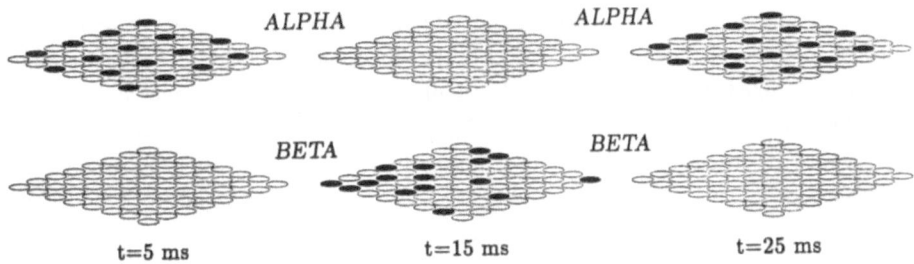

ALPHA ALPHA

BETA BETA

t=5 ms   t=15 ms   t=25 ms

Figure 5. This oblique view reveals the arrangement of the layer-IV neuronal assemblies in the modelled cortical areas ALPHA and BETA. As can be seen from the fact that just 16 of the assemblies are activated (black colour), the overall activity level of the input pattern was 25%. It should be noted that there is no geometrical significance to the activity pattern appearing in BETA. That area merely serves as a sort of sounding board for the coherent signals impinging upon it via the forward projections from ALPHA.

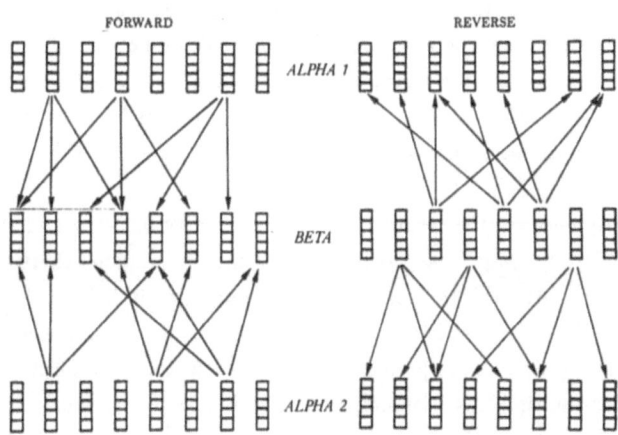

FORWARD   REVERSE

ALPHA 1

BETA

ALPHA 2

Figure 6. The situation in which two different input areas, designated ALPHA 1 and ALPHA 2, both interact with a higher cortical area BETA, via forward and reverse projections. This illustration should be compared with that shown in Fig. 4, and it should be noted that the pattern of activity produced in BETA will be the result of correlations jointly established by the coherent features of the inputs to ALPHA 1 and ALPHA 2.

areas. It is thus natural that one should attempt to extend the above simulations to include further areas, and we have in fact taken things one step further. As illustrated in Fig.6, we studied the situation in which two different input areas, designated ALPHA 1 and ALPHA 2 both engaged in forward and reverse projections with a higher area BETA. As before, the situation is illustrated by imagining the different areas to be arranged as if they were parallel to one another, since this is topologically equivalent to the actual situation.

The situation corresponding to that depicted in Fig. 5 will now be one in which there are simultaneously inputs to both ALPHA 1 and ALPHA 2, and the resulting pattern in BETA, which will of course be established about 10-15 ms later, will be the joint product of the coherent features in both inputs. Such a distributed pattern of active minicolumns corresponds well to the ideas proposed by Swindale[19]. And because of the return of (distributed) signals via the reverse projections, there will again be the possibility of generating renewed activity in the input areas, as indicated in Fig. 7.

Nothing has been stated as yet concerning memory effects, and a fuller discussion of that issue will be reserved for the discussion section of this paper. But it should here be remarked that the time lapse between the original injection of activity (from the dorsal lateral geniculate nucleus) to areas ALPHA and the renewed generation of activity in those areas, as mediated by the reverse projections, will be between 20 ms and 25 ms. It is interesting to note that this duration is comparable to the rise time constant[20] for the multiplicative activation of receptors belonging to the NMDA class. Such receptors, having been activated, then relax back to their initial state with a time constant of the order of 170 ms[20].

This similarity between the return time for the activity patterns and the activation time for the NMDA receptor[21] has suggested a mechanism whereby short term memory could be established. And this possibility must be considered in light of the fact that the distribution of NMDA receptors is not uniform throughout the various cortical layers. On the contrary, these receptors are preferentially located in the superficial cortical layers, whereas their non-multiplicative counterparts, the kainate and quisqualate receptors are most strongly represented in the deep layers[22]. Recalling what was stated earlier regarding the synapsing of the reverse projections onto the distal dendrites of the layer-IV neurons, those processes having their extremities up in layer I, we believe that the multiplicative learning mediated by the NMDA receptors may take place only in the reverse-projection paths.

We incorporated such learning in our simulational model, by imposing a 25 ms learning window, and we adopted a learning rate of 1 per cent per pass. The learning criterion was assumed to be of the Hebbian type, which is to say that a synapse was strengthened only if both the presynaptic and the post-synaptic neurons were active within the assumed

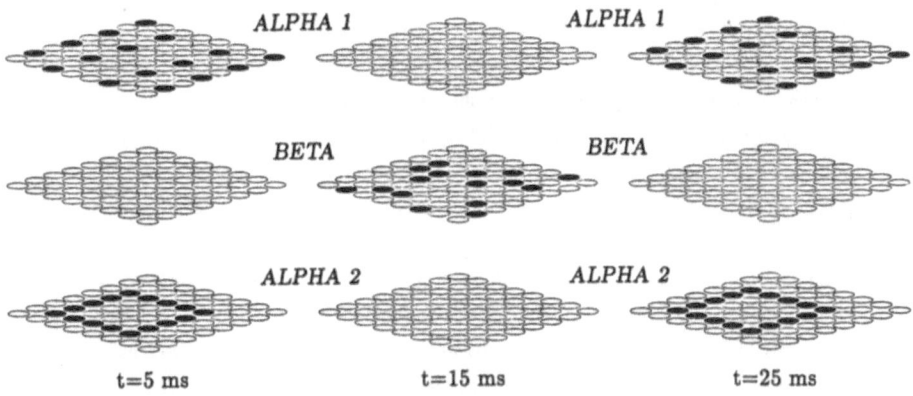

Figure 7. As in Fig. 5, an oblique view reveals the activity patterns in the various areas ALPHA 1, ALPHA 2 and BETA, and they are shown at 5 ms intervals as before. It is seen that the activity pattern developed in BETA subsequently produces patterns in ALPHA 1 and ALPHA 2 that closely resemble those originally injected into those areas, the return signals having been passed via the reverse projections.

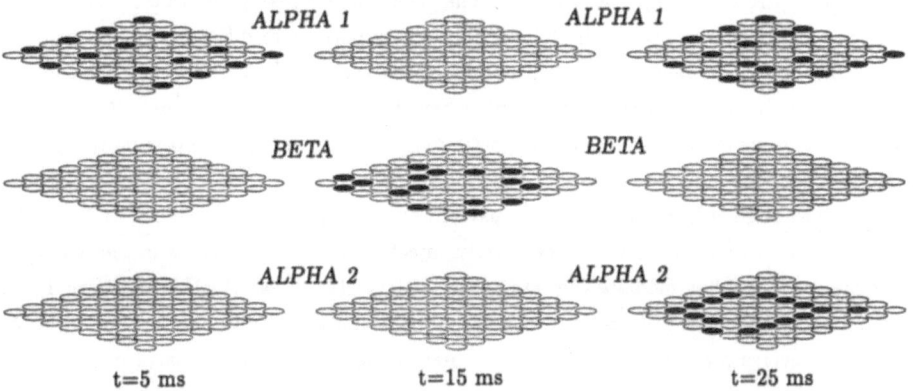

Figure 8. Following presentation of (different) patterns at the two input areas ALPHA 1 and ALPHA 2 (see also Fig. 7), Hebbian learning, exclusively in the reverse- projection routes, was implemented at the rate of one per cent per pass. This learning process was continued for one hundred round-trip passes of the system, which would occupy just two seconds at the 50 Hz frequency of the observed gamma oscillations. As can be seen from the figure, subsequent presentation of only one of the originally injected patterns, at the appropriate area, led to re-generation of the other pattern in the area where it was originally injected, and the re-generation process sets up oscillations between areas that are not directly stimulated by the sensory input.

NMDA activation window. The key feature of this activation in the case of our modelled cortex was that activity was, in the favorable cases, being returned to neurons whose membranes had been depolarized within the previous 25 ms.

After one hundred passes of such Hebbian learning (which would occupy a mere 2 seconds at 50 Hz), exclusively involving the reverse projections, we investigated what happened in our double-input (three-area) system following injection of only one of the original two input patterns, in the appropriate area. We observed that this led to the reappearance of the other original input pattern, in the area where it had originally been injected. The system had thus learned how to associate the patterns. This recall process is illustrated in Fig. 8. It is interesting to note that this regeneration mechanism would also involve the re-establishment of oscillations between area BETA and the area that is not receiving direct sensory input. It would be interesting to investigate experimentally the possibility of such remote generation of cortical oscillations.

DISCUSSION

We have not yet considered the pattern re-generation processes at the ALPHA areas in sufficient detail, and we will now proceed to do this in the context of the simple two-area system illustrated in Fig. 4 and Fig. 5. That the original pattern will be re-established in that area does not necessarily follow, and it is important to differentiate between the possibilities of such re-generation with and without the aid of a learning process, such as the one involving the NMDA receptors that have been invoked in this paper. We will first present arguments which support the possibility of re-generation of an originally injected pattern even when no such learning takes place.

As has already been discussed, the pattern of activity generated in area BETA is determined by two factors, namely the pattern injected into area ALPHA and the strengths of the synapses that mediate the interaction of layer II+III of ALPHA and layer IV of BETA, via the forward projections. Similarly, the pattern of activity returned to ALPHA, via the reverse projections, will be jointly determined by the actual pattern of activity in BETA and the strengths of the reverse-projection synapses. In general, of course, it will not be the case that those minicolumns in ALPHA that receive the strongest return signals from BETA will be identical with those that received the originally injected signals from the dorsal lateral geniculate nucleus.

Whether or not a given minicolumn in ALPHA will be activated after the above discussed approximately 21 ms will be determined both by what it receives from the reverse projections and what it might still be receiving from the lateral geniculate nucleus. And it will be the latter factor that will be particularly important, because it seems reasonable to assume that the strengths of the reverse-projection synapses will all be roughly equal, initially at least. It will thus be the minicolumns that are still receiving input that will be the most favoured ones. And given the lateral inhibition that is known to be in effect, those minicolumns that are not benefitting from such LGN input will lose out in the competition with their more highly activated neighbours. This is a manifestation of that frequently encountered principal: winner-take-all.

The originally-activated minicolumns having been reactivated, at the expense of their less well endowed neighbours, the process is ready to start over again, and the oscillation is thereby established, its frequency being determined by the ALPHA-BETA-ALPHA round-trip time. In effect, the oscillations can be regarded as a form of reverberation and as such they will constitute a type of very short memory. This would, however, be a rather different type of memory than that usually associated with the term because it would involve no changes at the synaptic level, neither chemical nor physical ones. Just how long the reverberation would continue is not easy to guess at, but there would presumably be a gradual diminution of the signal strength due to various loss factors.

There appears to be no recognized term that conveys the special "no-modification" quality of this form of memory, but we would like to propose that the designation "ephemeral memory" best describes it. Such ephemeral memory will, we feel, be an important aspect of consciousness itself because without it our senses would suffer from the drawback usually described by the term "tunnel vision", though senses other than the visual may of course also be involved. It is important to the correct functioning of any sensory modality that it can reap the advantage of the continuity provided by such ephemeral memory.

Ephemeral memory will give rise to short-term memory if the conditions are such as to lead to actual modifications at the synaptic level, and we have already noted in this paper that the NMDA receptors will provide the basis for such modifications. We have also already seen that one manifestation of this short-term memory will be the storage of associations, a phenomenon that certainly could not be observed on the basis of unaided ephemeral memory.

One aspect of consciousness having become involved in this discussion, this will be a convenient point at which to consider some other facets of this important property of the brain. Another reasonably obvious prerequisite of consciousness will be what is frequently referred to as binding[23]. This refers to the need of the organism to know which stimuli in its environment stem from one and the same source, irrespective of which sensory modalities are involved in its observation. A simple case in point is provided when one claps one's hands in front of one. The sense of touch informs the brain that the hands have come together, while the senses of hearing and vision simultaneously inform the brain of the same thing. But by what agency is the brain informed of such simultaneity, when the three different types of signal are arriving at different parts of the brain? There is no internal time-keeper dutifully keeping track of the various comings and goings. But comings and goings are precisely what are mediated by the forward and reverse projections! And by permitting the various senses to interact, in the reverberatory fashion discussed above, it is they that allow the brain to bind together sensory inputs that are causally related.

The third and final aspect of consciousness that appears to find a relatively straight-forward explanation through the observed gamma oscillations is the ability to discern coherent features in a sensory "scene", and this has already been explained, in effect, by the analysis given earlier of the manner in which area BETA serves its intermediary role. This appears to be the extent of what the observations allow one to state regarding consciousness, and it is worth emphasizing, at this point, that the possibility of self-consciousness makes further demands of the organism, demands that may be met only in sufficiently advanced species. Just what these additional requirements might be are naturally a matter of conjecture, but it seems reasonable to conclude that they include a cortex having a sufficient degree of complexity. Complexity, in this connection, might refer to what is present in any cortical area and it might also refer to the actual number of such areas. Indeed, one measure of a species' evolutionary position might be the number of other cortical areas that a given area can react with.

This brings us back to an interesting detail of Fig. 3 that has not yet been discussed in this paper, namely the fact that the periodic activity patterns observed in the areas ALPHA and BETA are mutually out of phase with each other. In this paper, we have avoided identifying area ALPHA with V1 and area BETA with V2 because it is observed experimentally that the oscillations detected in V1 and V2 are approximately in phase. We prefer to look upon area ALPHA as a composite input area, and we are encouraged in this view by the fact that both area V1 and area V2 do in fact receive sensory input from the dorsal lateral geniculate nucleus.

If this is a tenable attitude, it would lead to the conclusion that our area BETA lies further up the hierarchy of visual areas, and its antiphase characteristic with respect to area(s) ALPHA would then become the object of considerable interest. Indeed, this would seem to present our findings with a possibly definitive test. Certainly, it would be intriguing if such antiphase behaviour could be detected experimentally. And if that were the case, another even more fascinating issue would be raised. For if it were the case that higher areas were not in phase with those lying earlier in a particular sensory pathway, would it perhaps be possible to think in terms of phase maps, with the various areas that lay at equivalent levels of the hierarchy displaying in-phase oscillations. This prospect is so exciting that there is an understandable temptation to hope that it is founded in reality.

References

1. Gray, C. M., & Singer, W. *Soc. Neurosci. Abstr.* **404**, 3 (1987).

2. Eckhorn, R., Bauer, R., Jordan, W., Brosch, M., Kruse, W., Munk, M. & Reitboeck, H. J. *Biol. Cybern.* **60**, 121-130 (1988).

3. Gray, C. M., Konig, P., Engel, A. K. & Singer, W. *Nature* **338**, 334-337 (1989).

4. Singer, W. *Concepts in Neurosci.* **1**, 1-26 (1990).

5. Crick, F. & Koch, C. *Seminars in the Neurosciences* **2**, 263-275 (1990)

6. Barinaga, M. *Science* **249**, 856-858 (1990)

7. Gilbert, C. D. & Wiesel, T. N. *Vision Research* **25**, 365-374 (1985).

8. Wiesel, T. N., & Gilbert, C. D. *Quart. J. exp. Physiol.* **68**, 525-543 (1983).

9. White, E. L. *Cortical Circuits* Birkhauser, Boston (1989).

10. Maunsell, J. H. R. & van Essen, D. C. *J. Neurosci.* **3**, 2563-2586 (1983).

11. Pandya, D. N. & Yeterian, E. H. in *The Cerebral Cortex* (eds Peters, A. & Jones, E. G.) **4**, 3-61 Plenum, London (1985).

12. Rockel, A.J., Hiorns, R.W. & Powell, T.P.S. *Brain* **103**, 221-244 (1980).

13. Mountcastle, V. B. in *The Neurosciences Fourth Study Programme* (eds Schmitt, F. O. & Worden, F. G.) 21-42 MIT Press, Cambridge, Mass. (1979).

14. Szentagothai, J. *Rev. Physiol. Biochem. Pharmacol.* **98**, 11-61 (1983).

15. Eccles, J.C.*Neuroscience* **6**, 1839-1855 (1981).

16. Martin, K. A. C. in *The Cerebral Cortex* (eds Peters, A. & Jones, E. G.) **2**, 241-285 Plenum, London (1984).

17. Jack, J.J.B., Noble, D. & Tsien, R.W. *Electic current flow in Excitable Cells.* Clarendon Press, Oxford, 177 (1975).

18. Douglas, R.J., Martin, K.A.C. & Whitteridge, D. *Neural Computation* **1**, 480-488 (1989)

19. Swindale, N. V. *Trends in Neuroscience* **13**, 487-492 (1990)

20. MacDermott, A. B. & Dale, N *Trends in Neuroscience* **10**, 280-284 (1987).

21. Mayer, M. L., Westbrook, G. L. & Guthrie, P. B. *Nature* **309**, 261-264 (1984).

22. Cotman, C. W., Monaghan, D. T., Ottersen, O. P. & Storm-Mathisen, J. *Trends in Neuroscience* **10**, 273-280 (1987).

23. Damasio, A.R. *Neural Computation* **1**, 123-132 (1989).

Modelling of cardiac rhythm:

from single cells to massive networks

D. Noble*

J.C. Denyer*

H.F. Brown*

R. Winslow+

A. Kimball+

*University Laboratory of Physiology
Parks Road
Oxford, OX1 3PT, UK

and

+Department of Physiology and Army High Performance Computer Center
University of Minnesota, USA

Abstract

The development of models of the sinoatrial node cell is reviewed to show the steady progress from the early models based on voltage clamp of small multicellular preparations to permeabilized patch clamp of isolated single cells. It is shown that the kinetics of the hyperpolarizing–activated current, i_f, are finely tuned to 'buffer' the cardiac pacemaker frequency against changes in other conductance parameters, in particular background conductances.

These single cell models have now been incorporated into large scale network models: N x N meshes with neighbouring cells electrically coupled by resistors representing gap junctions. With 128 x 128 sinus cells, and random distribution of intrinsic properties, only two nexus channels (unit conductance 50 pS) between neighbouring cells are required to entrain the cells and between 20 and 200 channels will synchronize activity almost completely throughout the array. These very low densities of gap junctions are consistent with experimental observations (Masson-Pevet et al, 1979). When intrinsic properties are distributed between the node centre and periphery using a Gaussian function, and with the same magnitude of cell-to-cell coupling, an excitatory wave starts in the peripheral regions of the node and propagates towards the centre, i.e. the *opposite* direction to that in the normal heart. This occurs experimentally when the rabbit sinus node is separated from the atrium (Kirchhoff et al, 1987). Surrounding the node model with an atrial network shifts the origin of excitation towards the centre of the node.

1 INTRODUCTION

This article is a progress report on work being done in collaboration between the Oxford cardiac electrophysiology group investigating the properties of isolated single cardiac cells and the University of Minnesota department of Physiology and the US Army High Performance Computer Center which are using massively parallel computers to model very large networks of electrically coupled cells. The first problem we have chosen to tackle is the mammalian sinoatrial node. There are two reasons for this choice. The first is that the experimental data obtained from cells in this area of the heart permits very accurate modelling to be performed, including the variation of properties known to occur between the different anatomical regions of the node. The second is that the mammalian sinus node, e.g. that of the rabbit, contains at most 200 000 – 400 000 cells (considering the central 0.1 mm^2 has been estimated to contain 5000 cells (Bleeker et al., 1980)) which is well within the range of networks that can now be realistically set up on parallel machines. We will first review the development of work on the single cell modelling and then report on progress being made in incorporating these models into massive networks.

2 SINGLE CELL MODELLING

Since Noma and Irisawa (1976) first succeeded in voltage clamping very small dissected preparations from the mammalian sinoatrial node a very detailed picture

has been developed of the conductance mechanisms involved in the natural pacemaker of the heart (for a recent review see Irisawa, Brown and Giles 1991). The
early experiments were very difficult to perform since the successful penetration
of such small cells with two microelectrodes was not easy. Moreover, it was impossible to tell what proportion of the cells in the multicellular preparation were
reasonably normal and what proportion had suffered from the dissection procedure. Even though it was likely that damaged cells simply disconnected themselves
from the remaining intact network (via calcium-induced closure of the nexus channels), so leaving those cells that were connected to the clamp electrodes reasonably
normal, this problem made it difficult to extrapolate from the properties of the
small network to the properties of individual cells. We will refer later in this article
to an important calculation (Noble, 1982) that is particularly dependent on this
kind of extrapolation.

It was therefore of prime importance, when the technique of single isolated
cells became available (see Noble and Powell, 1987, for reviews of these techniques
and their results), to determine whether the method of enzyme digestion of the
connective tissue could be used on the tiniest cardiac cells of all: the sinus node
cells. The first results were reported by Tanaguchi, Kokubun, Noma and Irisawa
(1981 — for other references to early work see Denyer and Brown, 1990) using
cells that seemed to have normal electrophysiological properties but which were,
however, abnormal anatomically: their appearance was round rather than spindle
shape. Isolated single cells tend to round-up when calcium balance is disturbed,
so this suggests both that the cells may have been, at least partially, calcium over-
loaded and, since their rhythmic properties were not obviously very different from
the intact tissue, that the rhythmic mechanism did not depend very strongly on internal calcium or on other parameters that may vary depending on the conditions
of the isolation method. An alternative explanation, demonstrated later in this
article, is that the sinus node pacemaker mechanism is *designed* to be particularly
resistant to conductance changes.

DiFrancesco, Ferroni, Mazzanti and Tromba (1986) succeeded in avoiding cell
rounding and obtained what they called 'spider' cells, with several protuberances.
Finally, Denyer and Brown (1988) succeeded in isolating single SA node cells which
retain their natural thin spindle shape (Masson–Pévet, 1979). A recent series of
papers (Denyer and Brown, 1990a, b, c) has used this method to describe both the

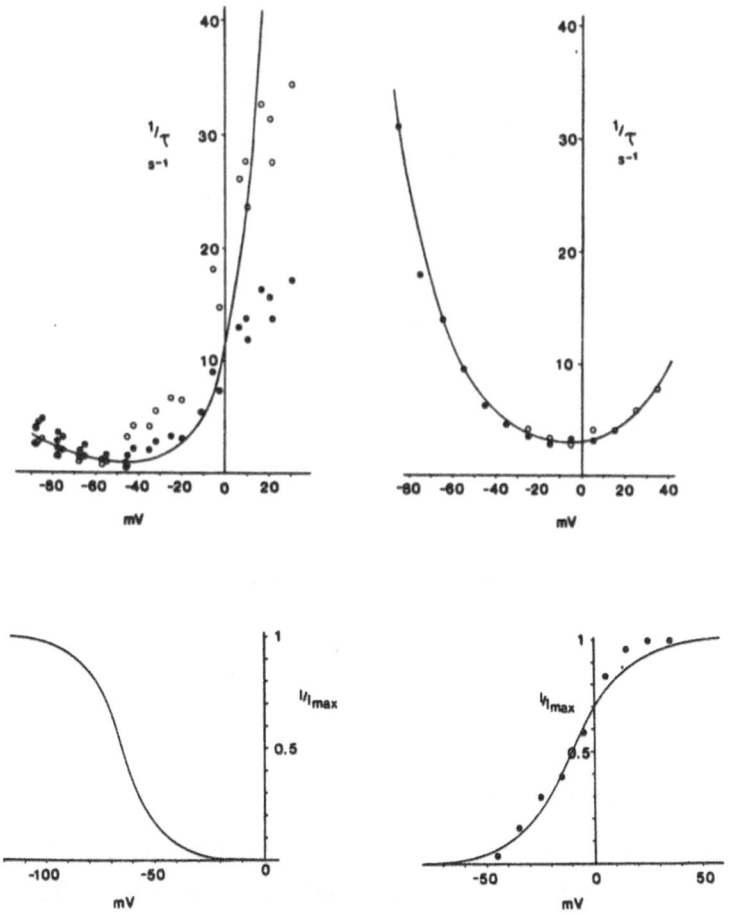

Figure 1. Kinetics of activation of i_K and i_f used in single sinus cell model. Left: the top diagram shows the reciprocal time constants of current change for i_f. The continuous line is computed from the functions used in the model. The bottom diagram shows the function used for the activation curve (see Figure 2 for the experimental results on which this is based). Right: Similar results for i_K.

basic electrophysiological properties and some of the mechanisms underlying rhythmic activity, including an attempt to assess the relative quantitative contributions of different inward currents to the generation of the pacemaker depolarization.

Nevertheless, a severe problem remained: this is that the patch clamping of

134

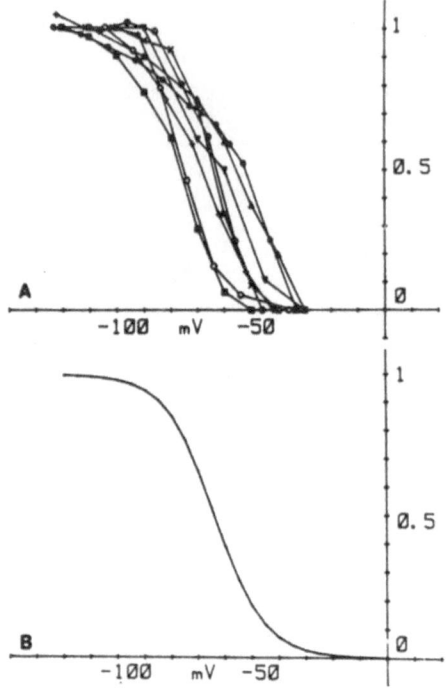

Figure 2. Top: activation curve for i_f in eight cells (different symbols refer to different cells). Notice the large variation in the position of the activation curve. Bottom: Theoretical function used as mean of the experimental results.

such small cells inevitably leads to rapid run-down of conductance mechanisms. The calcium channels progressively disappear and the hyperpolarizing-activated current i_f shifts to progressively more negative activation ranges. This is the explanation for the wide variation in the activation range measured for this current (see, for example, DiFrancesco and Noble, 1989, Figure 2, in which the i_f activation curve varies over a range of about 35 mV in the position of the voltage for half activation). This is a particularly severe problem when it comes to assessing the contribution of i_f to the pacemaker depolarization since this depends entirely on the position of the activation range. Even a shift of 10 mV in this range can lead to a change in contribution from this current mechanism of more than 100%.

Very recently, a method has been developed that permits patch electrodes to be used for clamping the whole cell without breaking the patch seal. Introduced by Horn and Marty (1988) it uses nystatin in the patch electrode to generate sufficient nystatin-induced conductance in the patch membrane to access the cell interior electrically without allowing the internal cell contents to be dialysed by the patch electrode.

This is the method that has been used to obtain the experimental data on which the latest computer model of single sinus node cells has been constructed (Noble, Denyer, Brown and DiFrancesco, 1991).

There are now, therefore, several versions of a mathematical model of the SA node using the approach first adopted by DiFrancesco and Noble (1985), including the original multicellular version (Noble and Noble, 1984), its extension to a single cell by Noble, DiFrancesco and Denyer (1989). There is also a very early model developed by Yanagihara, Noma and Irisawa (1980) which was the first SA node model, but this does not include concentration changes. We have extensively studied the properties of these models as a prelude to using them in very large network models and some important general conclusions have emerged.

2.1 Internal calcium

A key feature of the approach introduced by DiFrancesco and Noble (1985) was that variations in internal (and external) ion concentrations must be reproduced as well as the membrane current. The original justification for this requirement was that it was otherwise impossible to map the new work using the i_f pacemaker current onto the older work using the i_{K2} hypothesis. This mapping was very successful (DiFrancesco and Noble, 1982) and it depended entirely on the ability to reconstruct variations in extracellular potassium and the activity of the sodium-potassium exchange pump. It was this success that encouraged them to extend the approach to include internal calcium by incorporating SR uptake and release using a simple version of a model based on Fabiato's (1983) calcium-induced calcium release hypothesis. Later, Hilgemann and Noble (1987) refined the approach by totally revising the SR equations and by incorporating cytosol calcium buffers, including troponin and calmodulin.

The sinus node models have retained these attempts to reconstruct internal calcium changes. It is important to ask therefore whether the reconstruction of normal rhythm depends in any significant way on which of these models for $[\text{Ca}]_i$

is used. We have checked this question by replacing the DiFrancesco-Noble SR modelling in the latest single SA node cell model by the Hilgemann-Noble SR and cytosol buffer equations. The result is very reassuring: provided that the calcium transient is made sufficiently rapid to reach its peak and decay before the end of repolarization (which is the case in rabbit atrial cells, for example — see Earm, Ho and So, 1990; Earm and Noble, 1990), the overall computed voltage changes do not depend very strongly on which of these internal calcium formulations is chosen. This is supported by the observation that the spontaneous activity recorded in SA node cells with EGTA included the patch pipette solution (DiFrancesco et al., 1986; Hagiwara et al., 1988) does not differ significantly from that recorded in the absence of EGTA (Denyer & Brown, 1990a,b; van Ginneken, 1987). An important conclusion here is that the role of calcium-dependent currents, such as the sodium-calcium exchange, must be fairly minimal in determining SA node pacemaker rhythm. The hypothesis, for example, that this is the current that provides most of the background inward current in SA node cells must be incorrect. It remains though to determine more precisely what role this current does play in these cells, even if it is relatively minor with regard to pacemaker rhythm.

2.2 Relative roles of i_f and inward background currents

This question is both important and still not fully resolved. It is also the question that has strongly motivated further refinements of the SA node cell models. The problem can best be presented by referring to figure 1 which shows some of the experimental data on which the Noble–DiFrancesco–Denyer model was based. The data is shown as symbols while the mathematical functions used are represented by filled lines. The data on the delayed K current is clearly fitted very accurately indeed, whereas the data on the hyperpolarizing-activated current, i_f, is fitted poorly. This is in part because the activation of this current is not strictly exponential (DiFrancesco, 1984), but the more important difficulty arises from the fact that the position of the activation curve varies between experiments (Figure 2). As noted already above, this range can be up to 30 mV in patch clamped cells. This is partly what motivated the recent experiments using the nystatin patch method. In the Noble–DiFrancesco–Denyer model we simply used the mean position of the activation curve.

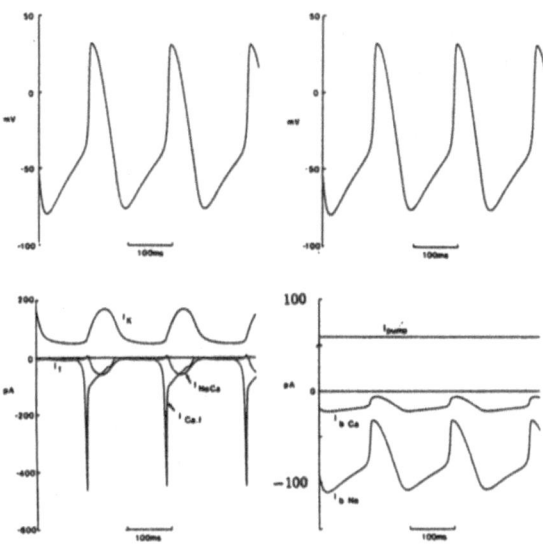

Figure 3. Computed pacemaker activity and ionic current using the single SA node cell model. Top: time course of computed voltage changes. Bottom left: time course of gated currents. Bottom right: time course of computed changes in background and pump currents.

Figure 3 shows the resulting computed pacemaker activity together with the variations in ionic currents underlying this activity. It can be seen that the activation of i_f is fairly small: its peak is only −25 pA compared to an estimated −100 pA for the background current. In this model therefore $i_{b,Na}$ contributes 4 times as much depolarizing current as does i_f. This is reflected in the fact that, when i_f is reduced to zero in the model (Figure 4) the change in frequency is relatively small. This is also observed experimentally (Denyer and Brown, 1990b; and Figure 5) when i_f is blocked with caesium ions.

It might be thought that this combination of experimental and computed results settles the issue. Figure 6 shows why this would be a premature conclusion. In this figure we have performed the converse "experiment", i.e. we have progressively blocked the background current, $i_{b,Na}$, instead of blocking i_f. Since this current carries 75% of the depolarizing current during the computed pacemaker

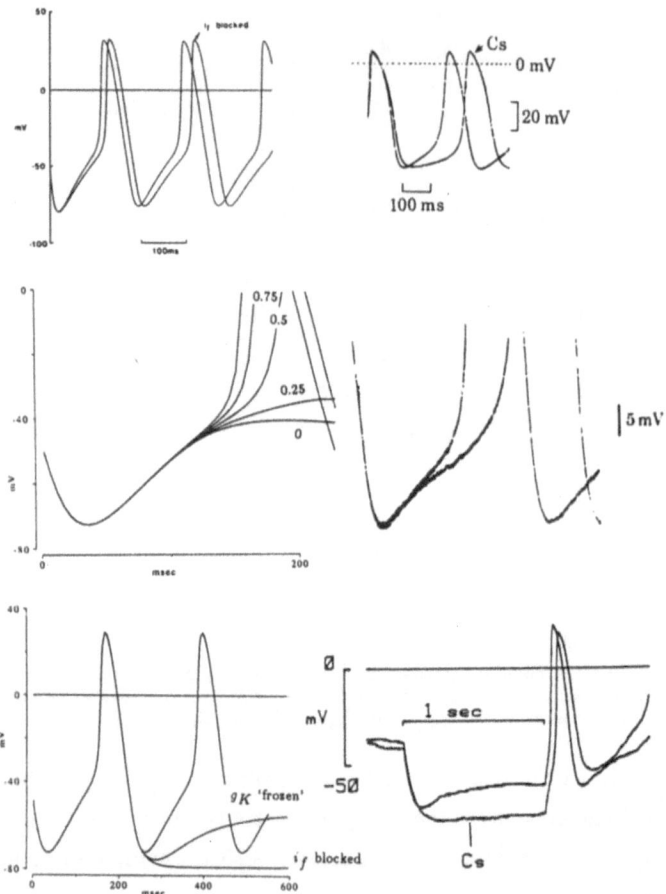

Figure 4. Top left: Effect of block of i_f on computed pacemaker activity in the single SA node cell model. Top right: example of an experiment in which a moderate slowing is induced in a single sinus node cell by blocking i_f with Cs ions. Middle left: influence of various degrees of block of calcium current on computed pacemaker depolarization. This slows the later phase of depolarization. Middle right: example of an experiment in which such slowing is achieved by vagal stimulation which is thought to reduce i_{Ca}. Bottom left: effect of stopping ('freezing') decay of g_K at the start of the pacemaker depolarization. This gives a very slow pacemaker depolarization which is completely removed by blocking i_f. Bottom right: Experiment in which a 'slow' pacemaker depolarization is recorded after applying hyperpolarizing current. This depolarization is also removed by blocking i_f with Cs ions.

depolarization it might be anticipated that the changes in frequency on changing its amplitude would be massive. In fact, however, the changes are not a lot larger than those obtained by blocking i_f. The top set of curves shows pacemaker activity computed from the model using background conductance values equal to 1,

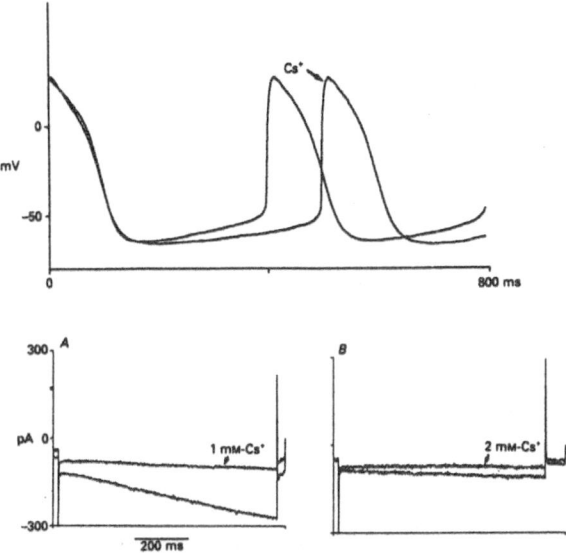

Figure 5. Top: whole cell nystatin-permeabilized patch clamp recording of spontaneous activity in an isolated SA node cell before and during superfusion with solution containing 2mM CsCl. Bottom: Influence of caesium on the hyperpolarization-activated current. The voltage clamp pulse is to −70 mV from a holding potential of −40 mV.

0.5, 0.25 and 0.125 of the normal value. There is always an initial delay due to the immediate hyperpolarization that occurs. After this delay, the rhythm settles down to frequencies only moderately smaller than the control level. Even when the conductance is reduced massively to 0.125 of its normal value the frequency change is no greater than that observed with complete block of i_f in many cells. Only by reducing $i_{b,Na}$ further does a dramatic change occur: pacemaker activity then stops.

The lower half of Figure 6 shows that even this assymetry between the two mechanisms cannot be taken as certain. Here we have repeated the calculations with the i_f activation curve shifted by 20 mV. This is within the range of experimental variation using the standard patch technique. It is clear that now the variation in frequency is extremely small and there was no level of conductance

140

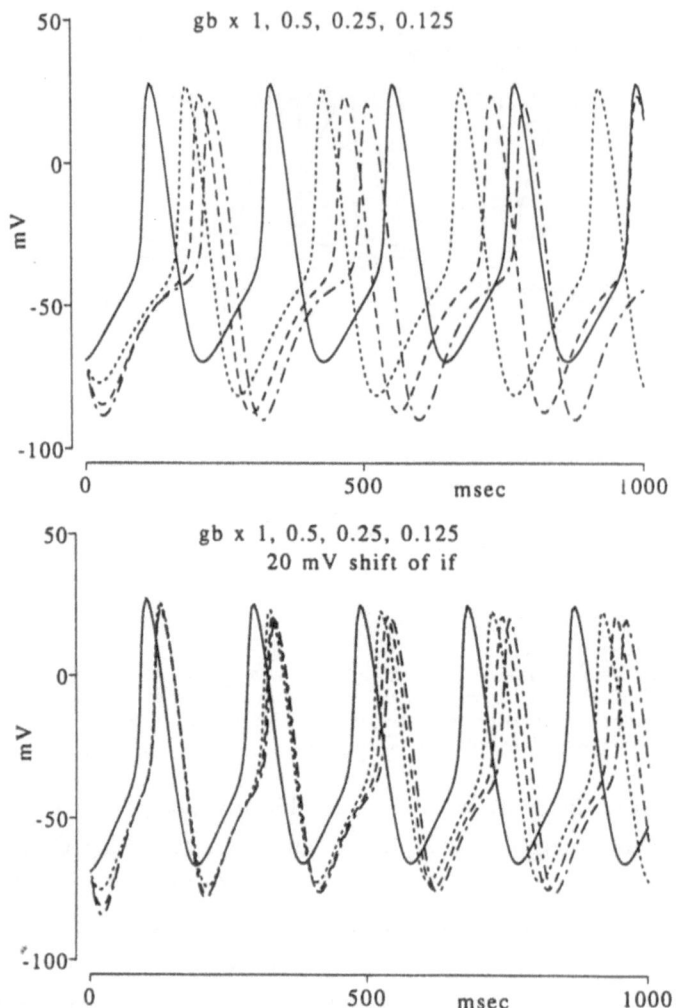

Figure 6. Top: Effect of reducing inward background current in SA node model to 0.5, 0.25, 0.125 of normal value. These changes induce moderate changes in frequency. Bottom: same computation repeated with i_f activation curve shifted 20 mV. The changes in frequency are now very small even when the background conductance change is very large.

(even zero) which would stop the rhythm. Moreover, this is very similar to the results obtained using the newer model (Noble, Denyer, Brown and DiFrancesco, 1991) based on the nystatin patch method, where the frequency shows a strong resistance to massive changes in background conductance and does not stop when the conductance is reduced to zero.

These results are surprising to say the least. What they mean is, first, that *either* i_f or $i_{b,Na}$ are fully capable, alone, of providing all the depolarizing current during the pacemaker potential up to the point at which the calcium current is activated. Second, they show that, regardless of the quantitative contribution that i_f actually makes to the normal depolarization (and this might vary between different regions of the SA node and in different biochemical circumstances), its kinetic properties are finely tuned for it to act as a buffer against frequency changes induced by other changes in conductance. This means that the sinus node pacemaker is extremely robust and resistant to factors that might arrest it. The survival importance of such a 'fail–safe' system is obvious.

3 NETWORK MODELLING

Although there are still some details to be developed, the models of single cells are sufficiently accurate to enable us to have some confidence in the idea that they could now form the basis of the ultimate goal of trying to put the heart back together again. Having taken the reductive pathway for the last 2 decades, with the progressive move from multicellular to single cell preparations, from net current clamping to single channel recording, from electrophysiology to its biochemical control, it is time to start reaping the rewards for our understanding of how the heart works as an organ. We have therefore incorporated the models of the single rabbit sinus node cell described above and the atrial cell model recently developed by Earm and Noble (1990) into large scale network models. These models are represented as N x N meshes with neighbouring cells electrically coupled by resistors representing gap junctions. Regional variation of intrinsic cell properties (as seen in central, transitional, and peripheral cells) such as frequency, amplitudes of action potentials and maximum diastolic potentials, upstroke velocity, and sensitivity to external [K$^+$] is included by adjusting the cell models to fit the data of Kodama and Boyett (1985). This information was obtained by segmenting a strip of the rabbit sinoatrial node running from its centre to the periphery and then recording the electrophysiological characteristics in each isolated segment. We have found it possible by fine tuning of the calcium and potassium conductances to reproduce Kodama and Boyett's data fairly accurately.

In the parallel computer computations, networks with as many as 1024 x 1024 cells are partitioned into sub-grids, mapping the set of differential equations defining each sub-grid onto different processors of a 32,768 processor Connection Machine CM-2 massively parallel processor (Thinking Machines Corporation), and integrating the equations for each sub-grid concurrently.

Our first question is designed to determine the degree of cell–to–cell connection required to enable the sinus node to act as a synchronous unit and to send a co-ordinated signal to the atrium. This question was last tackled by one of us in 1982 (Noble, 1982) when a calculation based on using the multicellular voltage clamp data suggested that sinus node cells with different intrinsic properties could be synchronised with only a very few nexus channel connections between them. This calculation suffered however from the assumption already referred to above, i.e. that the great majority of the cells in the multicellular preparations were behaving normally. The result of this assumption was that the single cell conductances were found to be exceedingly small so that even a pA of current flowing between two cells could have a significant effect on their rhythm. If, however, only a small fraction of the cells in the multicellular preparations were acting normally then the calculated individual cell conductances would have been much larger since the observed total preparation conductance would have been divided into fewer units. Since we now know the single cell conductances very accurately the question can now be answered definitively.

To achieve this we have arranged 128 x 128 sinus cells into a network on the Connection Machine, and, initially, we have assumed random distribution of intrinsic properties. The nexus conductance between cells was then progressively increased from 2 channels to 2000 channels (of 50 pS each) between cell neighbours. The results (Winslow, Kimball, Noble and Denyer, 1991; Winslow, Kimball, Varghese and Noble, 1992) show that as few as two nexus channels connecting neighbouring cells can achieve considerable frequency entrainment, while 20 channels achieve almost complete synchrony. The original conclusion of Noble's (1982) calculation is still therefore valid. These figures correspond to a very low density of gap junctions, which would occupy a very small fraction indeed of the cell membrane surface. This result is consistent with the experimental observations (Masson-Pevet et al, 1979) showing that in the central region of the sinus node nexus regions are indeed very sparse and occupy less than 0.2% of the cell surface.

The second question we have tackled with the network modelling is how the wave of excitation propagates when the intrinsic single cell properties (fitted to the Kodama and Boyett data) are distributed between the node centre and periphery using a Gaussian function. With the same magnitude of cell-to-cell coupling as in the random network, the results show that an excitatory wave then starts in the peripheral regions of the node and propagates towards the centre, i.e. in the *opposite* direction to that in the normal heart. This result is very encouraging since this is exactly what occurs experimentally when the rabbit sinus node is separated from the atrium (Kirchhoff et al, 1987). Moreover, when the simulated sinus node is surrounded by an atrial network the site of origin of the impulse shifts towards the centre of the node (Winslow et al, 1992). The network modelling is therefore already achieving significant results in line with experimental evidence. Our intention in the future is to extend the modelling to arrthythmic mechanisms so that we can study the way in which cellular and multicellular mechanisms of arrhythmia interact.

Acknowledgements: Supported by the Army High Performance Computing Research Center, The University of Minesota, Minneapolis, MN, Thinking Machines Corporation, Cambridge, MA, The Northeast Parallel Architecture Center, Syracuse University, Syracuse, NY, the BHF, MRC and Wellcome Trust. An earlier version of this paper was also presented at the 18th International Symposium of the Korean Academy of Medical Sciences held in Seoul on 19th February 1991

REFERENCES

Bleeker, WM, MacKaay, AJC, Masson-Pévet, M, Bouman, LN and Becker, AE (1980) Functional and morphological organization of the rabbit sinus node. *Circulation Research* **46** 11-22.

Denyer JC and Brown HF (1987) A method for isolating rabbit sinoatrial node cells which maintains their natural shape. *Japanese Journal of Physiology* **37** 963–965.

Denyer JC and Brown HF (1990a) Rabbit sinoatrial node cells: isolation and electrophysiological properties. *Journal of Physiology* **428** 405–424.

Denyer JC and Brown HF (1990b) Pacemaking in rabbit isolated sino-atrial node cells during Cs^+ block of the hyperpolarizing-activated current i_f. *Journal of Physiology* **429** 401–409.

Denyer JC and Brown HF (1990c) Calcium 'window' current in rabbit isolated sino-atrial node cells. *Journal of Physiology* **429** 21P

DiFrancesco D (1984) Characterisation of the pacemaker current kinetics in calf Purkinje fibres. *Journal of Physiology* **348** 341–367.

DiFrancesco D, Ferroni, A, Mazzanti M and Tromba C (1986) Properties of the hyperpolarizing-activated current (I_f) in cells isolated from the rabbit sino-atrial node. *Journal of Physiology* **377** 61–88.

DiFrancesco D and Noble D (1982) Implications of the re-interpretation of i_{K2} for the modelling of the electrical activity of pacemaker tissues in the heart. In: *Cardiac Rate and Rhythm* Ed. Bouman,L.N. & Jongsma, H.J. pp.93–128. The Hague, Boston, London: Martinus Nijhoff

DiFrancesco D and Noble D (1985) A model of cardiac electrical activity incorporating ionic pumps and concentration changes. *Philosophical Transactions of the Royal Society* **B 307** 353–398.

DiFrancesco D and Noble D (1989) The current i_f and its contribution to cardiac pacemaking. In: *Cellular and Neuronal Oscillators* (Ed. JW Jacklet) 31–57, New York: Dekker

Earm YE, Ho WK and So IS (1990) Inward current generated by Na–Ca exchange during the action potential in single atrial cells of the rabbit. *Proceedings of The Royal Society* **B 240** 61—81.

Earm, Y. E. and Noble, D. (1990) A model of the single atrial cell: relation between calcium current and calcium release. *Proceedings of The Royal Society* **240**, 83–96.

Fabiato A (1983) Calcium-induced release of calcium from the cardiac sarcoplasmic reticulum. *American Journal of Physiology* **245** C1–C14.

Hagiwara, N, Irisawa, H and Kameyama, M (1988) Contribution of two types of calcium current to the pacemaker potentials of rabbit sinoatrial node cells. *Journal of Physiology* **395** 233-253.

Hilgemann DW and Noble D (1987) Excitation–contraction coupling and extracellular calcium transients in rabbit atrium: Reconstruction of basic cellular mechanisms. *Proceedings of the Royal Society* **B 230** 163–205.

Horn R and Marty A (1988) Muscarinic activation of ionic currents measured by a new whole-cell recording method. *Journal of General Physiology* **92** 145–159.

Irisawa, H, Brown HF & Giles WR (1991) Cardiac pacemaking in the sinoatrial node. *Physiological Reviews* (in press)

Kirchhoff CJHJ (1989) *The sinus node and atrial fibrillation*. Ph. D. thesis, University of Maastricht.

Kirchhoff CJHJ, Bonke FIM, Allessie MA and Lammers WJEP (1987) The influence of the atrial myocardium on impulse formation in the rabbit sinus node. *Pflügers Arch* **410** 198–203

Kodama, I. and Boyett, M. R. (1985) Regional differences in the electrical activity of the rabbit sinus node. *Pflügers Arch.* **404**, 214–226.

Masson-Pevet, M., Bleeker, W. K., Mackaay, A. J. C., and Bouman, L. N. (1979) Sinus node and atrium cells from the rabbit heart: a quantitative electron microscopic description after electrophysiological localisation. *J. Molec. Cell. Cardiol.* **11**, 555–568.

Noble D. (1982) In discussion following De Haan, RL. (1972) In: Cardiac Rate and Rhythm ed. Bouman, LN & Jongsma, H.J. pp359–361. Martinus Nijhoff: The Hague.

Noble D, Denyer JC, Brown HF and DiFrancesco D (1991) Modulation of pacemaker rhythm by conductance changes. (in preparation).

Noble D, DiFrancesco D and Denyer JC (1989) Ionic mechanisms in normal and abnormal cardiac pacemaker activity. *Neuronal and Cellular Oscillators* J. W. Jacklet, Dekker: New York, 59–85

Noble D and Noble SJ (1984) A model of S.A. node electrical activity using a modification of the DiFrancesco-Noble (1984) equations. *Proceedings of the Royal Society* **B 222** 295–304.

Noble D and Powell T (1987) *Electrophysiology of single cardiac cells*. Academic Press.

Noma A, and Irisawa, H. (1976) Membrane currents in rabbit sinoatrial node cells studied by the double microelectrode method. *Pflügers Archiv* **364** 45–52.

Tanaguchi J, Kokubun S, Noma A and Irisawa H (1981) Spontaneously active cells isolated from the sino-atrial and atrio-ventricular nodes of the rabbit heart. *Japanese Journal of Physiology* **31** 547–558.

van Ginneken, AGC (1987) Membrane currents in mammalian cardiac pacemaker cells. Ph. D. thesis. Amsterdam University.

Winslow R, Kimball A, Noble D, Denyer JC and Varghese A (1991) Simulation of very large sinus node and atrial cell networks on the Connection Machine CM-2 massively parallel computer. *Journal of Physiology* **438** 180P.

Winslow R L, Kimball A L, Varghese A and Noble D (1992) Simulating cardiac sinus and atrial network dynamics on the connection machine. *Physica D: Nonlinear phenomena*, in press.

Yanagihara K, Noma A and Irisawa H (1980) Reconstruction of sino-atrial node pacemaker potential based on the voltage clamp experiments. *Japanese Journal of Physiology* **30** 841–857.